Weeds and Poisonous Plants of Wyoming and Utah

Edited By

Tom D. Whitson

Photography

Tom D. Whitson
University of Wyoming
Extension Weed Specialist

Roy Reichenbach
Supervisor, Converse County
Weed and Pest Control District

Text Contributors

Tom D. Whitson
University of Wyoming
Extension Weed Specialist

Mark A. Ferrell
University of Wyoming
Extension Pesticide Coordinator

Stephen D. Miller
University of Wyoming
Associate Professor of Weed Science

Steven A. Dewey
Utah State University
Extension Weed Specialist

John O. Evans
Utah State University
Professor of Weed Science

Richard J. Shaw
Utah State University
Emeritus
Professor of Botany and Director of
Intermountain Herbarium

April 1987

TEACHING • RESEARCH • EXTENSION
UW
EXCELLENCE
Cooperative Extension Service
College of Agriculture
The University of Wyoming

Issued in furtherance of Cooperative Extension Work, acts of May 8 and June 30, 1914, in cooperation with the U.S. Department of Agriculture. F.E. Busby, Director, Cooperative Extension Service, University of Wyoming 82071.

A cooperative publication of the University of Wyoming, Cooperative Extension Service and Utah State University Cooperative Extension Service and Agricultural Experiment Station.

Library of Congress Catalogue Card Number: 87-50423
ISBN: 0-941570-06-1

Table of Contents

Introduction

Species discussed in this publication represent only a fraction of the weedy or potentially troublesome plants found in Utah and Wyoming. An attempt has been made to include a representative cross-section of the more common agronomic, range and non-crop weeds in the two states.

There are many ways to define a weed. We feel the following most accurately describes the criteria used for including plants in this book:

A plant that interferes with management objectives for a given area of land at a given point in time.

-- J. M. Torell

The term "weed" does not always indicate that a plant is totally undesirable, or that it cannot be beneficial under certain situations. According to the above definition, many desirable species might sometimes be considered weedy. For example, a species providing valuable wildlife forage or wildlife habitat on one tract of land may be considered undesirable on land managed to maximize grass production for livestock. It all depends on the management objectives. Some plants poisonous to livestock are considered desirable ornamental plants. Other species that are almost always considered undesirable may actually provide limited benefit as they rapidly invade disturbed sites and thereby reduce soil erosion.

Some of the weeds listed in this publication may not yet have been introduced into Utah and/or Wyoming. However, their potential for invasion and establishment warrants concern. Any infestations of these weeds should be reported as soon as they appear so eradication efforts may be initiated.

To promote standardization of common plant names, this publication uses common and scientific names from the Weed Science Society of America's official list of common and scientific names. Family concepts are based on the 1980 Cronquist system.

Because of frequently changing weed control recommendations, none have been given in this publication. For current control measures, check with your County Cooperative Extension Service or Weed and Pest Office.

Acknowledgement

Appreciation is extended to faculties of the Rocky Mountain Herbarium, University of Wyoming, and the Intermountain Herbarium, Utah State University, for their assistance in confirming plant identification for this publication.

Appreciation is also extended to the University of Wyoming Division of Communication Services for their assistance in the design and printing of this publication: Debra Beck, Publications Editor; Katherine Bella, typesetting; Elizabeth Ono Rahel, graphic design; Carol Hale Stevens, graphic design.

Weeds and Poisonous Plants of Wyoming and Utah

Western brackenfern
Pteridium aquilinum (L.) Kuhn var. *pubescens* Underw.

Western brackenfern

Polypodiaceae (Dennstaedtiaceae)
(Fern family)

Coarse medium-sized to large (1.5 - 6 feet tall) perennial ferns with branched, creeping, woody rhizomes. Fronds are leathery, pinnate to decompound and densly hairy on the underside. Brown spores are borne along the undersurface edge of each frond segment, and are protected by a narrowly inrolled indusial leaf margin. Fronds turn brown and die in autumn. Newly emerging unrolled fronds appear as fiddlenecks in the spring. *Pteridium* is distinguished from other fern genera primarily by the fact that its rhizomes have no scales. Other genera will have scales, or scales and hairs on rhizomes.

Brackenfern is a native species found throughout most of the western U.S. It is most likely to occur in open woods or mountains where soil pH is neutral or acidic. Brackenfern is not generally considered an aggressive invading weed, but it is included for its potential to cause livestock poisoning. Animals accustomed to its presence rarely graze fern, but stock newly introduced to infested areas may consume fatal quantities. Poisoning in horses and sheep appears to be cumulative, and symptoms may not be evident until some time after initial feeding.

Fronds, leaves of ferns, have a knobby appearance in early spring as they develop.

Compound fronds underneath side, hairy developing reddish-brown reproductive spores in late summer.

Scouringrush
Equisetum hyemale (L.)

Scouringrush

Equisetaceae
(Horsetail family)

A tenacious perennial plant with deep, spreading rootstocks. Stems are rough (high silica content), jointed and hollow. Leaves are scale-like appendages located at stem nodes. Fertile stems have a terminal spore-producing cone. Several species of *Equisetum* are found in the intermountain West, including field horsetail (*Equisetum arvense* L.), and scouring rush (*Equisetum hymale* L.). Field horsetail is a dimorphic species. Some plants having simple stems, and others producing whorls of branches at each stem node. Scouring rush produces only simple stems.

Field horsetail is adapted to wet areas, while scouring rush can be found either along streambanks and or in relatively dry soils. *Equisetum* can create serious maintenance problems along highway rights-of-way and irrigation waterways; and is difficult to control in cropland. At least some species within this genus are poisonous to livestock, affecting primarily horses and cattle.

Reproductive spore producing cones develop in July and August.

Rigid hollow stems are segmented and are easily pulled apart at nodes.

Common cattail
Typha latifolia (L.)

Common Cattail
Typhaceae
(Cattail family)

Tall perennial herbs with large creeping rhizomes. Stems are simple, cylindrical, erect and pithy. The parallel-veined linear leaves are long, broad, flat, and rather spongy. Mature inflorescences are velvety-brown, densly crowded, terminal, spike-like, cylindrical heads. These eventually fall apart, releasing clouds of tiny tufted, air-borne fruits.

Cattail is commonly found in wet or marshy places throughout the intermountian region, where it provides beneficial habitat for waterfowl and other wildlife. Though cattail is not generally considered a weed, its presence in water storage ponds or along irrigation canals and ditches can reduce holding capacity or impede waterflow.

Large creeping rhizomes readily propagate new shoots, allowing dense stands to develop.

Velvety-brown seed heads break apart releasing many air-borne fruits.

Stinging nettle
Urtica dioica (L.)

Stinging nettle
Urticaceae
(Nettle family)

A perennial slow-spreading plant; 2 - 9 feet tall; with four angled stems; coarsely-toothed, opposite, paired leaves; and numerous, small, bristly, stinging hairs over much of its surface. Greenish flower clusters are of two types, with staminate and pistillate flowers usually on different plants. Clusters are borne on slender branches in axils of upper leaves.

Stinging nettle is a native species confined primarily to shaded, moist areas along streams; or in deep, rich, undisturbed soils. It is primarily a nuisance to recreationists because of its stinging hairs. It also produces impenetrable, unmanageable stands along waterways used for livestock or irrigation purposes.

Leaves of stinging nettle are coarsely-toothed, forming opposite pairs. Stems are square. Undersides of leaves and stems have small needle-like structures.

Green flower clusters are attached to the axil of the leaf and stem.

Erect knotweed
Polygonum erectum (L.)

Prostrate knotweed and Erect knotweed

Polygonaceae
(Buckwheat family)

Extremely variable annuals with wiry stems. Nodes are somewhat swollen, often surrounded with the torn remnant of a short, papery leaf sheath. Leaves are generally lance-shaped, blue-green and alternate. Inconspicuous, greenish-white flowers are borne in the leaf axils. Several species occur in our region, including prostrate and erect knotweed. *Polygonum aviculare (L.)* branches basally becomes prostrate, while *Polygonum erectum (L.)* is more upright.

Knotweed is a native of the Old World, but is now distributed throughout North America. It can become a problem in irrigated and dryland crops, as well as in waste areas and other non-crop sites. Prostrate knotweed seems to thrive in dry compacted soils.

Prostrate knotweed lies flat on the soil surface.

Inconspicuous greenish-white flowers were magnified against a cloth background.

Wild buckwheat
Polygonum convolvulus (L.)

Wild buckweat
Polygonaceae
(Buckwheat family)

An annual weed with stems trailing along the ground, often twining about other plants. Leaves are heart-shaped with a tapered point, and are arranged alternately along the stem. An inconspicuous papery sheath encircles the stem at the base of each leaf petiole. Clusters of tiny greenish flowers are borne in leaf axils. Seeds are triangular, dull black, slightly roughened and about 1/2 inch long.

Wild buckwheat was introduced from Europe, and is now a common weed in cultivated fields, gardens, orchards and non-crop areas of our region. Superficially resembling field bindweed, wild buckwheat can be distinguished by its annual habit, pointed leaves, papery leaf sheaths and small green flowers. (Bindweed flowers are white to lavender, large, showy and trumpet-shaped.)

Seedling plants have distinctive heart-shaped leaves.

Clusters of greenish flowers are borne in leaf axils. Each flower contains a hard, triangular black seed.

Pale smartweed
Polygonum lapathifolium (L.)

Pale smartweed and Ladysthumb

Polygonaceae
(Buckwheat family)

Annual weeds, with erect or spreading stems usually 1 - 3 feet long. Stems may initiate roots at lower nodes. leaves are alternate, narrow, lance-shaped, often marked with a characteristic pink or purplish "water spot" near the middle. Leaf nodes are conspicuous and sheathed, as is common with most members of this genus. Flowers are small, pink or rose-colored; they are borne in dense, erect, terminal and axillary spikes approximately 1 inch long. Pale smartweed and ladysthumb, *Polygonum persicaria (L.)* are similar in appearance, but may be differentiated by the fact that ladysthumb sheaths are tipped with short bristles, while those of smartweed lack bristles. Smartweed spikes tend to be longer, and the flowers a lighter pink. Ladysthumb seeds are shiny, black, and may be either flattened or three-angled. Pale smartweed seeds are flattened, shiny and brown.

These amphibious species are often found in wet or moist undisturbed sites, but they can also become troublesome in cultivated fields and irrigated pastures. Several other species also occur in our region.

Ladysthumb leaves contain a darkened area near the center of the leaves. Smartweed leaves are consistant in color.

Smartweed spikes of flowers tend to be longer and lighter pink than ladysthumb (not pictured).

Curley dock
Rumex crispus (L.)

Curley dock
Polygonaceae
(Buckwheat family)

A robust tap-rooted perennial growing 2 - 5 feet tall. Stems are erect, generally unbranched on the lower half, often reddish and slightly ridged. Leaves are elongated (4 - 12 inches long) and lack hairs. They have curly or wavy margins. Flowers are small and in dense, green, spike-like, terminal and axillary clusters. Inflorescences and even entire plants turn reddish-brown at maturity. Individual seeds are enclosed in a papery, sometimes corky, winged structure (1/8 -3/16 inch long) that facilitates distribution by wind or water.

Curley dock is a native of Europe that has become well established in our region. It is especially common in wet meadows, along ditch banks and in waste places. It can also become troublesome in grain fields, pastures and lawns. Several other related species also occur, including broad-leaved dock (*Rumex obtusifolius*), spinach dock (*Rumex patientia*), sheep sorrel (*Rumex acetosella*) and veiny dock (*Rumex venosus*).

Rosettes in early spring have twisted leaves 4 - 12 inches long with wavy margins.

Seeds are individually contained within winged structures aiding in their dispersal.

Common lambsquarters
Chenopodium album (L.)

Common lambsquarters
Chenopodiaceae
(Goosefoot family)

Extremely variable annual; 1 - 6 feet tall, stems erect, much branched and often striped with pink or purple, leaves alternate, the lower often wavy margined to somewhat lobed, the upper narrower and often entire, undersurface white to grayish-mealy. Flowers are small, inconspicuous, gray-greenish mealy, crowded in axils and at tips and branches of stem, seeds numerous, black and shiny.

A native of Europe, lambsquarters has become established throughout most of North America. It is common in cultivated fields, gardens, and waste areas. It is a very competitive weed because of its rapid growth and high water use. Lambsquarters also serves as a host for the beet leafhopper which transmits the curly top virus to sugarbeets. Lambsquarters is often eaten in salad or as greens when the plant is young and tender. Flowering and seed production may occur from July to September.

Seedling with leaves showing the goosefoot shape with tiny silver glands on the leaf surface.

Flowers are inconspicuous and appear as a mealy substance in leaf axils and the terminal stem.

Halogeton
Halogeton glomeratus (Stephen ex Bieb.) C.A. Mey.

Halogeton
Chenopodiaceae
(Goosefoot family)

An annual weed ranging in height from a few inches to over 18 inches. Main stems branch from the base, spreading at first, and then becoming erect. Plants are blue-green in the spring and early summer, turning red or yellow by late summer. Leaves are small, fleshy, and nearly tubular, ending abruptly, tipped with a delicate needle-like spine. Flowers are green and inconspicuous, borne in leaf axils. Plants resemble Russian thistle in early stages. But they can be distinguished by leaf shape, and by the presence of minute cottony hairs in the leaf axils.

Halogeton is a native of Asia that has rapidly invaded over 11 million acres in Utah, Wyoming, and other western states. It seems ideally adapted to the alkaline soils and semi-arid environment of our high-desert winter livestock ranges. Halogeton is not an extremely competitive plant, but it readily invades disturbed or over-grazed lands. It is usually most concentrated along roadsides, sheep trails and near areas where livestock congregate. The plant produces toxic oxalates that are especially poisonous to sheep, though cattle may also be affected. Halogeton is readily grazed at times, and is believed responsible for thousands of livestock poisonings.

Fleshy leaves terminating with a needle-like structure connect individually to a red colored stem.

Green flowers borne in leaf axils in the fall surround each plant stem.

Kochia
Kochia scoparia (L.) Schrad

Kochia
Chenopodiaceae
(Goosefoot family)

Annual, 1 - 6 feet tall, stems much branched, round, slender, usually soft-hairy, but occasionally smooth, often red-tinged; leaves alternate, lance shaped, entire 1/2 to 2 inches long, margins fringed with hairs. The upper surface is usually smooth, the under surface usually covered with soft hairs; leaf blades with three or five prominent veins. The flowers are inconspicuous, sessile in the axils of upper leaves and form short, dense, bracted spikes; seed wedge-shaped, dull brown, slightly ribbed and approximately 1/16 inch long.

Native of Asia introduced from Europe, kochia has escaped from cultivation as an ornamental and is now found throughout North America. It is common in cultivated fields, gardens, roadsides, ditchbanks and waste areas throughout the western U.S. While it is usually considered as an objectionable weed, kochia is readily grazed by livestock. Further, it has been reported that at early growth stages, kochia's nutritive value and palatability compared favorably with alfalfa. Flowering and seed production may occur from July to October.

Seedlings of kochia germinate in very early spring and should be treated with herbicides in this stage.

Flowers are inconspicuous, forming dense spikes in leaf axils.

Russian thistle
Salsola iberica (Sennen and Pau) also*Salsola kali* (L.) var. *ruthenica* (Iljin) Soo and var. *tenuifolia* Tausch.

Russian thistle
Chenopodiaceae
(Goosefoot family)

Also referred to as *Salsola kali*, *Salsola kali* var. tenuifolia, *Salsola kali* var. ruthenica, and *Salsola pesitfer*. A rounded bushy annual, much branched, 1/2 - 3 feet tall, reproducing by seed. Stems are usually red or purple striped. Leaves are alternate; the first is long, string-like and soft, with later leaves short, scale-like and tipped with a stiff spine. Inconspicuous green flowers are borne in leaf axils of upper leaves, each flower accompanied by a pair of spiny bracts.

Seeds are spread as mature plants break off at ground level and are scattered by the wind as tumbleweeds. Rapid germination and seedling establishment occur after only brief and limited amounts of precipitation. Since Russian thistle was introduced (from Russia) in the late 1800s, it has become one of the most common and troublesome weeds in the drier regions of the U.S. It is well adapted to cultivated dryland agriculture, but is also found on disturbed wastelands, overgrazed rangeland, and even some irrigated cropland. Barbwire Russian thistle (*Salsola paulsenii* Litv.) is similar in overall appearance; but it is generally more coarse and robust, with broader and more rigid spine-tipped leaves.

Seedling plants have long, fleshy leaves; herbicide applications should be made at this growth stage or before.

Stems are striped with purple at most growth stages. Magnified flowers are found in leaf axils in the fall of the year.

Greasewood

Sarcobatus vermiculatus (Hook.) Torr.

Greasewood
Chenopodiaceae
(Goosefoot family)

A native perennial shrub becoming a dominant species on saline alkaline flood plains. This woody species has rigid stems which are white to gray in color growing over seven feet tall. Linear round leaves are pale green, usually less than one inch long and attached without a petiole to a woody stem. Flowers appearing late summer are yellow in color, inconspicuous with female flowers located below male flowers on the same plant. The plant starts growth in early spring and loses its leaves after frost.

Greasewood is a moderately poisonous plant that must be consumed in large amounts by sheep and cattle for death to occur. Sheep are often poisoned in the fall from eating large amounts of fallen leaves. Toxicity increases as the plants mature in the fall. Leaves contain oxalates of potassium and sodium which cause death of livestock in four to six hours after they have been eaten. Sheep may die after eating up to two pounds of leaves while cattle may consume three to four pounds for death to occur. Commonly used herbicides defoliate plants but they usually resprout causing greasewood to be a difficult species to control.

Young greasewood plants have fleshy leaves and are individually connected to white woody stems

Flowers appear inconspicuously in the fall with the leaves.

Tumble pigweed
Amaranthus albus (L.)

Tumble pigweed
Amaranthaceae
(Pigweed family)

An annual species, somewhat prostrate to spherical and erect, much branched, 6 - 36 inches tall. Leaves usually have narrower blades and shorter stalks than those of prostrate pigweed. Stems are usually light green rather than red, more erect, more intricately branched, and the general aspect of mature plants is more spiny than prostrate pigweed. Flowers are borne in small spiny clusters at the base of leaves, not in long terminal spikes as in the case of redroot pigweed. Seeds are black, shiny, lens-shaped and 0.7 to 0.8 millimeters wide.

Tumble pigweed is an introduced species found primarily in cultivated or disturbed sites. This plant is a prolific seed producer (10 - 12 million seed/plant) and the seed can be spread great distances when mature wind-blown plants break off and tumble along the ground.

Fruit are borne in leaf axils rather than on branch terminals.

Seed of all pigweeds appear about the size of a pin head and are shiny black.

Prostrate pigweed
Amaranthus blitoides (S. Wats.)

Prostrate pigweed
Amaranthaceae
(Pigweed family)

An annual with generally prostrate stems radiating in all directions from a central taproot. Main stems are usually 12 - 18 inches long with shorter secondary branches. All stems are somewhat fleshy and pliable, nearly smooth, and usually red to purple. Leaves are small and egg-shaped, with the tip-end broader than the base. Flowers are in small congested clusters in the leaf axils. Long terminal flower spikes are absent. Seeds are shiny, black, lens-shaped and approximately twice the width of tumble pigweed seeds.

Prostrate pigweed was probably introduced from tropical America, adapting well to our area. It occurs mostly in disturbed or cultivated soils, and is often associated with tumble pigweed or other Amaranthus species. It is a common garden weed.

Leaves of prostrate pigweed are oval shaped with red or purple stems. Flowers are contained in leaf axils.

Seedling plants also have red to purple stems with oval leaves.

Redroot pigweed
Amaranthus retroflexus (L.)

Redroot pigweed
Amaranthaceae
(Pigweed family)

A coarse erect annual, usually 2 - 3 feet tall. Lower stems are often red or red-striped, with color continuing down the taproot. Leaves have long petioles with prominent veins. They are somewhat broad and lance-shaped, and often become reddish. Individual flowers are small, green and tightly arranged in large, branched, spike-like, terminal clusters. Smaller axillary flower clusters may also occur. Flower clusters are full of stiff, spine-like scales, making this pigweed additionally undesirable in hay. Seeds are small, black and shiny.

Redroot pigweed was probably introduced from tropical America. It is widely distributed throughout our region, commonly found in cultivated lands, gardens, and waste areas. Pigweed normally begins growth in late spring, and matures in late summer or early fall. However, seeds may germinate anytime during the growing season when soil moisture is adequate. Significant crop yield reductions can be caused by competition from this large, fast-growing weed. Additionally, it is reported to cause livestock poisonings as a result of high nitrate levels.

Seedlings of redroot pigweed have distinct leaf veins and leaves are connected to stems with long petioles.

Terminal flower clusters are plume-like with stiff spine-like projections surrounding them. Thousands of shiny black seeds are borne in each structure.

Common Purslane
Portulaca oleracea (L.)

Common purslane
Portulacaceae
(Purslane family)

A fleshy, prostrate, annual weed with smooth reddish or flesh-colored stems. Branches radiating from a central rooting point reach lengths in excess of 12 inches and form dense vegetative mats. Smooth, shiny, succulent leaves are somewhat tear-drop shaped; wider at the tip than at the base. Five-petalled yellow flowers are borne singlely in leaf axils, and open only in sunshine. Numerous, tiny, black seeds are produced in capsules resembling flower buds.

Introduced from Europe, common purslane has become a troublesome weed in cultivated fields and gardens. It is especially persistent in soils that remain moist much of the time. Production of seed throughout the growing season, and the ability to take root again after cultivation make this plant especially difficult to control. Also, seeds can remain dormant in the soil for years before germinating. Purslane has some limited value as a potherb, but is rarely eaten.

Leaves and stems are fleshy with small yellow flowers appearing in leaf axils in late summer.

Seedlings of purslane have teardrop-shaped succulant leaves that are wider at the tip than at the base.

Common chickweed
Stellaria media (L.)

Common chickweed
Caryophyllaceae
(Pink family)

Annual or winter annual, reproducing by seeds and creeping stems, rooting at the nodes; roots shallow and fibrous. Stems are sometimes ascending but commonly prostrate forming mats of plants 4 -12 inches high, numerous branches, round in cross-section with a conspicuous line of hairs on one side. Leaves are 1 - 4 cm long, opposite, simple, ovate to oblong, entire, the upper leaves sessile, the lower stalked and often hairy toward base or on petiole. Flowers are 1/4 inch in cross-section, star-shaped, white, perfect, solitary in the axils or in cymose clusters; sepals 5, separate, 3-6 mm long, lanceolate-oblong, blunt to acute, usually hairy. Petals are white, deeply two-parted, shorter than sepals.

Common chickweed can be found throughout the world in habitats ranging from the tropics to high altitudes. The plant is a particularly troublesome weed since its seeds are able to germinate during the winter months under favorable conditions. The plant, introduced from Europe, is commonly found in gardens, strawberry beds, alfalfa and lawns where it can seriously impede the growth of desirable plant species.

Leaves are opposite with hairs present on the base of each leaf. Stems have a single line of hairs down them.

White perfect flowers are found in leaf axils or on stem terminals.

Tall larkspur
Delphinium occidentale (Wats.) Wats.

Tall larkspur
Ranunculaceae
(Buttercup family)

A perennial up to 5 feet tall; leaves are large, simple alternate, lobed into three to five divisions which are lobed again; root tuberous, woody. Flowers, purple, terminal, five petals with a distinct spur pointing backwards. Seeds are borne in an upright pod; stem hollow.

It is native to the Rocky Mountain region and may be distinguished from plains larkspur because it grows at higher elevations, is usually much taller, has larger leaves and stems, and flowers in the summer instead of spring. All parts of the plant including seeds contain toxic alkaloids. Fatal poisoning occurs mainly in cattle, whereas losses rarely occur in sheep and horses. Cattle losses may be minimized by keeping them off larkspur infested ranges until other forage is available, grazing the area with sheep before allowing cattle to graze, or by treating with herbicides. When using herbicides special precautions must be taken to keep cattle off treated areas, since applications of herbicides increase the palatability of larkspur.

Leaves are alternate on stems. Each leaf is lobed into sections then individual sections are lobed again.

Purple flowers contain five petals with a spur-like appendage pointing backwards. Each is connected to the stem by a petiole.

Geyer larkspur
Delphinium geyeri (Greene)

Geyer larkspur
Ranunculaceae
(Buttercup family)

Geyer Larkspur is a perennial which grows up to 2 feet tall; leaves are finely divided, lobed into three to five divisions which are lobed again; root tuberous, woody; flowers, purple, terminal, five petaled with a distinct spur pointing backwards; seeds borne in an upright pod; stem hollow.

Geyer larkspur is native to the Rocky Mountain region and is common on western rangelands in the plains and mountain areas. It is similar in appearance to tall larkspur but may be distinguished because of its smaller, more finely divided leaves. It flowers in the spring, grows in lower elevations, and is smaller in height. Toxic alkaloids cause fatal poisoning in cattle, whereas sheep and horses are rarely affected. It is rarely eaten after it reaches maturity in late June. Cattle losses may be reduced by keeping them off infested ranges until other forage is available, grazing with sheep before allowing cattle to graze, or by treating with herbicides. When using herbicides special precautions must be taken to keep cattle off treated areas, since application of herbicides increase the palatability of larkspur.

Early spring growth of Geyer larkspur is the most poisonous to cattle and is one of the first plants to grow on rangeland.

Flowers, connected to the main stem by petioles have five petals with a spur on their back sides.

Annual pricklepoppy
Argemone polyanthemos (Fedde) G.B. Ownbey

Annual prickleypoppy
Papaveraceae
(Poppy family)

Annual prickleypoppy grows up to 3 feet tall; reproduces by seed; leaves are grayish-green, coarsely lobed, spiney on margins and midrib; stems erect, prickley, sparingly branched. Flowers are terminal, large, white. Entire plant contains a yellow sap. Seeds are small, round, blackish-brown, ridged on one side.

Annual prickleypoppy is a native plant occuring in pastures, waste areas, and along roadsides. It is seldom abundant and can be an indicator of overgrazing. Its spines make it undesirable for livestock feed. The prickles have been known to irritate the skin of some people. Although it produces beautiful flowers, it is generally considered a nuisance because of its unpalatability.

Yellow spines along the main stem and leaves prevent livestock from using this plant.

Large showy flowers with yellow centers become apparent in late summer.

Rocky Mountain beeplant
Cleome serrulata (Pursh)

Rocky Mountain beeplant
Capparaceae
(Caper family)

Rocky Mountain beeplant is an annual which grows up to 3 feet tall; reproduces by seed; leaves three parted, alternate; flowers numerous, pink or white. Seeds are yellow-brown, 1/8 inch long, born in a slender pod 1 - 2 inches long.

Rocky Mountain beeplant is native to the western region and frequently occupies dry sandy soils along roadsides and waste places. The blossoms are filled with nectar making it attractive to bees. Indians used the plant for food and making pottery paint. It is undesirable as a forage for livestock due to its unpleasant odor. Because of its showy flower it is commonly found in gardens.

Immature plants have pink stems with three-parted leaves.

Flowers containing many showy stamen appear either white or pink. Seed pods are 1 - 2 inches long.

Smallseed falseflax
Camelina microcarpa (Andrz. ex DC.)

Smallseed falseflax
Brassicaceae
(Mustard family)

This is an annual; 1 - 3 feet in height; leaves alternate, simple, entire, auriculate-clasping at the base of the stem; pubescence either forked or stellate; inflorescence raceme; petals pale yellow; fruit a silicle, obovoid. There are several reddish-brown seeds in each chamber.

Smallseed falseflax was introduced from Europe. It is found on disturbed sites such as fields, roadsides and waste places. It is frequently brought into a new area with feeds. Smallseed falseflax spreads by seeds from late spring to early summer.

An immature plant shows the alternate leaf arrangement and entire leaves.

Reddish-brown seeds are contained in each of the oval capsules. Flower petals are pale yellow.

Shepherdspurse
Capsella bursa-pastoris (L.) Medic.

Shepherdspurse

Brassicaceae
(Mustard family)

Winter annual or annual, 3 - 18 inches tall; stems erect, slender, one to several from a basal rosette, hairy below and smooth above; leaves alternate with the lower ones usually deeply lobed, the upper ones are few, lance-shaped, toothed or entire, clasping the stem with a pair of ear-like lobes at the base. The flowers are white, small and usually found at the end of elongated racemes, fruits more or less heart-shaped, two-celled; seeds many, very small, oblong, reddish-brown with a single ridge on each side.

Native of Europe, shepherdspurse has become well-established throughout much of the U.S. and Canada. It is common in cultivated fields, gardens, waste areas, poorly maintained pastures and on roadsides. This plant is one of the first to green up in the spring and is relished by chickens. Flowering and seed production may occur from April to September.

Seedlings are deeply lobed, forming a rosette on this winter annual.

Fruit are heart-shaped and contain many reddish-brown seeds. Heart-shaped capsules have a ridge down each side and are attached by a long pedicel to the main stem.

Hoary Cress
Cardaria spp.

Hoary Cress
Brassicaceae
(Mustard family)

Four species of hoary cress are common in the western U.S. with differences in seed capsules and fruit used to identify each species. A deep rooted perennial up to 2 feet tall reproducing from root segments and seeds. Leaves are blue green in color, lance-shaped. Lower leaves are stalked; upper leaves have two lobes clasping the stem. Plants have many white flowers with four petals, seed capsules contain two heart-shaped, reddish-brown seeds separated by a narrow partition. Flowers are dense and give the plant a white, flat-topped appearance. Plants emerge in very early spring and have bloomed and set seed by mid-summer.

This perennial is common on alkaline, disturbed soils and is highly competitive with other species once it becomes established. It can be controlled effectively with herbicides in early growth stages.

This perennial starts growth very early in the spring—herbicides should be applied at this growth stage.

Many white flowers with four petals develop into bladder-like seed capsules in mid-summer.

Blue mustard
Chorispora tenella (Pall.) DC.

Blue Mustard

Brassicaceae
(Mustard family)

This is an annual; 5-20 inches high, leaves oblanceolate, simple, sessile, entire to deeply lobed; glandular pubescence, sticky to the touch; disagreeable odor; petals rose to purple; inflorescence a compound cyme; flowers perfect; siliques indehiscent and terete; seeds in a single row in each chamber.

Blue mustard was introduced from Asia. It grows in grain fields, waste places, roadsides and other disturbed places. This mustard is spreading faster than any other mustard in Utah. Blue mustard resembles African mustard (Malcomia africana R. Br.) but is distinguished by seed pods that break into two-seeded segments. Blue mustard spreads by seeds from early spring to early summer.

Seedling rosettes have deeply lobed leaves on this winter annual mustard.

This winter annual is one of the first plants to flower in early spring. Flowers have four opposite petals and are connected with a pedicel to a rough central stem.

Flixweed
Descurainia sophia (L.) Webb ex Prantl

Flixweed and Pinnate tansy mustard

Brassicaceae
(Mustard family)

Flixweed is an annual; 8 - 24 inches high; leaves alternate, two to three times pinnately compound; segments very narrow or linear; pubescent with branched hairs or nearly glabrous; inflorescence a raceme; petals very small, yellow or greenish-yellow, silique-septum with two to three longitudinal nerves, siliques 1/2 to 1 1/4 inches in length.

Flixweed was introduced from Europe. It is found growing in waste places, fields, roadsides and other disturbed sites. This mustard is distinguished from other tansy mustards because of its finely dissected leaves. Flixweed spreads by seeds from early to late summer.

Pinnate tansy mustard, *Descurainia pinnata* (Walt.) Britt is an annual, 4 - 32 inches high; leaves alternate, pinnately dissected; stellate pubescence; inflorescence a raceme; petals yellow or yellowish green to cream; siliques less than 3/4 inch long; siliques clavate or linear; two rows of seeds in each seed pod.

A native growing in waste places, fields, roadsides, and other disturbed sites. Tansy mustard spreads by seeds from early to late summer.

Tansy mustard seed capsules (siliques are less than 3/4 inch long. Petioles or stems holding seed capsules are near the same length as the capsule or longer.

Seedling plants of flixweed and tansymustard are very similar. Herbicides should be applied at this growth stage or before for effective control.

Dyers woad
Isatis tinctoria (L.)

Dyers woad
Brassicaceae
(Mustard family)

Dyers woad is a winter annual, biennial or short-lived perennial; 12 - 39 inches in height, leaves alternate, simple, basal, petiolate, bluish-green, whitish nerve on the upper surface of the blade; inflorescence an umbel; petals yellow; fruit a silicle, indehiscent, black or purplish brown and one celled, one seeded pod.

Dyers woad was introduced from Europe. It made its first appearance in the United States in colonial times. Dyers woad was first observed in Utah in 1917 and from there spread to Wyoming. This mustard has a thick tap root which may exceed 5 feet in depth. This characteristic allows the rosette to reproduce asexually from the tap root if it is removed. Dyers woad is first found along roadsides and railroad right-of-ways and spreads from there to rangeland and cropland. It spreads by seeds from late spring to mid-summer.

Seedling plants germinate in the fall and over winter in this stage. Herbicide applications are most effectively applied in this growth stage.

Purplish-brown seed pods containing a single seed appear near mid-summer, giving this weed a totally different look.

Perennial pepperweed
Lepidium latifolium (L.)

Perennial pepperweed

Brassicaceae
(Mustard family)

This plant is a perennial; 1 - 3 feet in height; leaves lanceolate, bright green to gray-green, entire to toothed, basal leaves larger than upper leaves; inflorescence raceme. Petals are white, in dense clusters near ends of branches, very small, fruit a silicle, seeds two per fruit, rounded, flattened, slightly hairy, about 1/16 inch long, and reddish-brown.

Perennial pepperweed, a native of Europe, is now found in many parts of the United States. In California, Utah and Washington it has been declared a noxious weed. Perennial pepperweed spreads both by seed and a deep-seated rootstock. The deep-seated rootstock makes it difficult to control. It is often found growing in waste places, wet areas, ditches and roadsides. Flowering occurs from early summer to early fall.

Leaves and stems are covered with a waxy layer making this perennial difficult to control.

Dense flower clusters appear in early summer on perennial pepperweed.

Clasping pepperweed
Lepidium perfoliatum (L.)

Clasping pepperweed
Brassicaceae
(Mustard family)

This plant is a winter annual or annual, 6 - 18 inches tall, stems erect, branched at the top and hairy; leaves alternate of two types, the lower dissected and the upper heart-shaped with a clasping base. The flowers are white to yellow with slender pedicels on racemes, fruit a two-valved obicular capsule, each containing two reddish-brown, somewhat rough, wing-margined seeds about 1/12 inch long.

Clasping pepperweed is a native of Europe and has become established in much of the western U.S. It grows scattered in grain fields, pastures and waste areas. Flowering and seed production occur from April to June.

Upper leaves are heart-shaped, clasping reddish-brown stems while basal leaves are finely divided.

Each of the oblong capsules contain two reddish-brown seeds. Seeds are formed in early summer from this winter annual.

Tumble mustard
Sisymbrium altissimum (L.)

Tumble mustard
Brassicaceae
(Mustard family)

Tumble mustard is a winter annual or annual, 2-5 feet tall; stems simple below, much branched above and bushy in appearance. Leaves alternate and of two types; lower leaves are coarse and divided into broad lobes or leaflets, the upper are much reduced, finer with narrow lobes or segments. The flowers are small, pale yellow in racemes, fruit a slender two-valved capsule 2-4 inches long. Seeds are small, numerous, yellow to brown, oblong. They usually have a single groove.

A native of Europe, tumble mustard is now widely scattered throughout the U.S. It is common in small grain fields, rangeland, waste areas and along roadsides. The plant often breaks off at the soil level when mature and scatters seed as it tumbles in the wind. Flowering and seed production occur from May to September.

A seedling rosette of this winter annual has leaves similar to the basal leaves of a mature plant, leaves are deeply lobed.

Light yellow, four-petal flowers are borne at upper stem terminals in early spring.

Field pennycress
Thlaspi arvense (L.)

Field pennycress
Brassicaceae
(Mustard family)

Field pennycress is an annual; 6 -18 inches high; basal leaves lanceolate, simple, entire to lobed, inflorescence raceme. Petals are white, pinkish or lavender, in clusters on the ends of branches; fruit a silicle, orbicular in outline, obviously winged, rounded, deeply notched at the top and seeds two or more in each chamber.

Field pennycress was introduced from Europe. It is a very troublesome weed in grain fields and is also found growing in waste places, roadsides and other disturbed areas. A strong odor is associated with this plant. It is frequently brought into a new area through feed. Field pennycress spreads by seeds and flowers from late spring to early summer. Dairy cows eating field pennycress produce a bitter-flavored milk.

Two seeds are contained in each chamber of this mature plant. They are surrounded by a fan-like structure.

Small white four-petal flowers are formed at the ends of field pennycress branches in early summer.

Twogrooved milkvetch
Astragalus bisulcatus (Hook.) Gray

Twogrooved milkvetch

Fabaceae
(Pea family)

Twogrooved milkvetch is a perennial upright herb growing to 30-inch heights. Flowers are generally purple, sometimes blue, cluster near the end of plant branches. Fruit is contained in a pod having two parallel rounded grooves. Opposite paired hairy leaflets, oblong in shape, vary in numbers but generally have from 11 - 30 per leaf. Stems are dark purple as plants mature. Plants are indicators of saline soils and have a strong selenium odor. Plants emerge in late spring and mature in early summer.

Many milkvetches are poisonous plants affecting cattle, sheep, and horses. Plants are poisonous throughout the growing season. As little as two pounds cause acute poisoning in mature cows within a few hours after being eaten. Plants contain glucosides which cause respiratory problems and paralysis of the hind legs in livestock. Animals often seek out plants, and many have become addicted to milkvetches once they graze them. Milkvetches can be controlled with herbicides before plants reach full bloom. Moving affected livestock to non-infested areas for long time periods will reduce losses.

Seed formation takes place in seed capsules with two grooves running parallel down each capsule. Stems have a purple color.

Purple or blue flowers cluster at terminal ends of branches in early summer.

Goatsrue
Galega officinalis (L.)

Goatsrue
Fabaceae
(Pea family)

Goatsrue is a perennial reproducing from seed and taproots often growing to heights over three feet. Leaves are odd-pinnate with five to eight pairs of leaflets. Flowers are purple, blue or white on terminal stems or in leaf axils. Stamens are united except the upper one which is partially free. Pods are narrow and round in cross-section slightly more than 1 inch long.

Goatsrue is an introduced perennial from the Middle East. This slow spreading weed occurs in varying degrees of infestation on about 38,000 acres in Cache County, Utah. Within this area it infests cropland, waterways, pastures, fencelines, roadways and wet, marshy areas. At maturity, goatsrue seeds drop to the ground and are primarily spread by irrigation water. They may be spread by harvesting equipment, animal manure or soil moving equipment. Although goatsrue is a legume and resembles alfalfa in some respects, it is very different in most agronomic characteristics. Goatsrue is not palatable to livestock, making it useless as a forage species. Therefore, it has a competitive advantage over most forage plants. This species can be controlled with herbicides but should be reported to your county agent or weed supervisor.

Seed pods are narrow, round in cross-section and slightly more than one inch long.

Flowers of goatsrue appear as blooms in axils of leaves or as terminal flowers on stems. Color is purple, blue or white.

Wild licorice
Glycyrrhiza lepidota (Pursh)

Wild licorice
Fabaceae
(Pea family)

Wild licorice is a perennial reproducing from deep, spreading roots or seeds; stems tall, 1 - 3 feet in height, erect, simple or with upper part producing erect branches; leaves alternate, pinnately compound with 11 - 19 deeply veined lanceolate leaflets, glandular dots when mature; flowers in short axillary spikes on long peduncles, calyx with two upper teeth shorter and partly united; corolla with narrow standard and blunt keel, green-white to white, stamens--nine fused by filaments and one separate; seed pod 0.4 - 0.75 inches long, burlike, covered with stout, hooked prickles; seeds 2 - 3 mm long, bean-shaped, reddish-brown, smooth and dull.

Wild licorice is a widely distributed native plant commonly found in moist, sandy soils of meadows, pastures, prairies, ditch and river banks and waste areas. The common licorice used to flavor candy is a different species, though the root of the wild licorice is equally sweet and was an important food source for the American Indian.

Alternate leaflets are pinnately compound with 11 - 19 lanceolate leaflets.

Burs less than 3/4 inch long appear in August and September on wild licorice

Wyeth lupine
Lupinus wyethii (S. Wats.)

Wyeth lupine

Fabaceae
(Pea family)

One of several lupine species in the western U.S. A perennial plant reproducing by seeds. Stems are upright and branched, often forming large showy clumps up to 18 inches in height. Flowers range from white to purple on Wyeth lupine which are more open than other species. Leaflets and stems are covered with fine hair. Palmate leaves are composed of six to eight leaflets radiating from a central point. Flowers mature from the bottom of the plant to the top forming hairy pods containing several round seeds.

Hungry sheep are often poisoned by lupine plants when being trailed through ranges in late summer. Cows eating lupines during early pregnancy sometimes have calves with skeletal defects. Lupines contain poisonous alkaloids throughout the growing season and even though there are some species that are not poisonous precautions should always be taken with livestock. Poisoning can occur in sheep ingesting less than 1/4 pound of the plants while cattle must eat over 1 pound for poisoning to occur. Control of lupine with herbicides should be done before the bud stage of growth.

Palmate leaves are composed of six or eight leaflets radiating from a central point (an excellent identification characteristic).

Flower color varies from white to blue and white on Wyeth lupine as well as other species of lupine.

Silky crazyweed
Oxytropis sericea (Nutt.)

Silky crazyweed and Lambert crazyweed

Fabaceae
(Pea family)

Silky crazyweed and Lambert crazyweed, *Oxytropis lambertii* (Pursh), are perennial herbaceous legumes common on western rangelands. Mature plants range in height from 6 - 12 inches. Leaves are covered with fine hairs giving a whitish-gray appearance. Flowers vary in color with species, with silky crazyweed having white flowers and Lambert crazyweed having purplish-pink flowers. Flowers are borne on a leafless seed stalk emerging from the center of the plant forming a spike-like cluster. The plant reproduces from kidney-shaped seed which are formed in a hairy, leather-like pod.

All livestock species can be poisoned by eating locoweed. Horses never recover once they are poisoned. Cattle gain weight slowly and often have abortions, while sheep abortions are high from eating these plants. Livestock generally avoid eating crazyweed until feed is scarce, but once they have eaten it, they seek the plants out. Herbicides should be applied to actively growing plants, but before they reach the bud stage.

Early growth stage of silky crazyweed is shown. Pinnately compound leaflets are covered with fine hair.

Lambert crazyweed flowers are purplish-pink and appear in mid-summer.

Swainsonpea
Sphaerophysa salsula (Pall.) DC.

Swainsonpea
Fabaceae
(Pea family)

Swainsonpea is a creeping perennial reproducing by seeds and lateral roots. Leaves are compound, composed of numerous opposite leaflets which are oval and covered with hair. Flowers occur from May to July, are 1/4 to 1 inch in length and are orange-red. Many seeds are formed in bladder-like translucent pods. Seeds are green with an attachment indentation.

This plant was introduced from Asia and is commonly found in western states along roadsides and fences. It is a potential threat to grain and alfalfa growing areas; the seed is a contaminant in alfalfa which is the same size, shape and weight making it impossible to separate.

Green seeds near the same size as alfalfa seeds can be seen through translucent bladder-like pods in mid-summer.

Orange-red showy flowers 1/4 - 1 inch long appear from May to July on swainsonpea.

Plains pricklypear
Opuntia polyacantha (Haw.)

Plains pricklypear
Cactaceae
(Cactus family)

Perennial forming low, spreading fleshy plants commonly in clumps but can be erect to 3 feet or more in height, reproducing from stems or seeds. Stems are flat and jointed, areoles 8 - 15 mm apart in spirals, spines 2 - 3 cm long, pale or brownish with around nine from an areole. Leaves are small and scale-like on young branches, dropping early. Flowers are large; calyx tube does not extend beyond ovary; petals numerous, 4 - 5 cm long, slightly united; stamens numerous in several rows, anthers 1.5 mm long. The fruit is a pear-shaped berry either juicy or dry, often spiny. Seeds are numerous, 4.5 - 6 mm long, white and flattened. Three species of *Opuntia* occur in Utah and Wyoming: *O. fragilis*, *O. rhodantha*, *O. polyacantha*. *O. rhodantha* and *O. polyancantha* both have flattened stems, though *O. rhodantha* has flattened spines and pink to red flowers. In contrast, *O. polyacantha*, the more common species of prickly pear, does not have flattened spines and its flowers are lemon yellow to orange.

Pricklypear is a native plant commonly found on dry, sandy soils. It can be a troublesome weed on overgrazed pastures and rangelands. Herbicide controls are effective on these species.

Individual plains prickly pear plants may be over three feet across with several blooms on the same plant.

Plainsprickly pear flowers varying in color from lemon yellow to orange and are waxy in appearance with a green stigma in the center.

Leafy spurge
Euphorbia esula (L.)

Leafy spurge
Euphorbiaceae
(Spurge family)

Perennial, up to 3 feet tall; reproduces by vigorous rootstalks and seed. Leaves are alternate, narrow, 1 - 4 inches long. Stems are thickly clustered. Flowers are yellowish-green, very small, arranged in numerous small clusters. They are enclosed by paired heart-shaped bracts also yellow-green in color. Roots are brown, containing numerous pink buds which may produce new shoots or roots. The entire plant contains a milky sap. Seeds are oblong, grayish to purple, contained in a three-celled capsule, each cell with a single seed.

Leafy spurge is native to Eurasia and was brought into the United States as a seed impurity about 1827. However, it seems to be a serious problem only in North American where it infests almost 2.5 million acres, mostly in southern Canada and the northcentral United States. It has also been reported to cause severe irritation of the mouth and digestive tract in cattle which may result in death. Capsules explode when dry, often projecting seeds as far away as 15 feet. Seeds may be viable in the soil for up to eight years. An extensive root system containing large nutrient reserves makes leafy spurge extremely difficult to control.

Heart-shaped yellow bracts surround the three-celled seed capsule, each cell containing a single seed.

Pink buds which form new shoots are common on leafy spurge crowns, and roots. Rooting depths of over 14 feet are reported with this prolific species.

Ridgeseed spurge
Euphorbia glytosperma (Engelm.)

Ridgeseed spurge

Euphorbiaceae
(Spurge family)

This annual, prostrate-growing plant forms mats, stems smooth, up to 15 inches long and much branched often reddish. Leaves are opposite, simple, oblong or linear oblong with small teeth toward the apex. The flowers are small in axillary clusters. Fruit is a glaborous three-angled capsule which is wider below the middle. Seeds are oblong, whitish, brown or nearly black, sharply four-angled with three to six transverse ridges passing though the angles.

Three species of prostrate milky spurge are common to Wyoming and Utah (ridgeseed, thyme leaf and hairy). The first two species are thought to be indigenous while hairy spurge has likely been introduced. All three species are common weeds of gravelly driveways, cultivated areas or in new or poorly established lawns. Flowering and seed production may occur from June to September.

Ridgeseed spurge seedlings germinate in early spring. Leaves are oblong and are individually attached to pink stems which contain latex milky sap.

Inconspicuous flowers and seed capsules are found in leaf axils in late summer.

Puncturevine
Tribulus terrestris (L.)

Puncturevine
Zygophyllaceae
(Calthrop family)

Puncturevine is annual, prostrate or somewhat ascending, mat forming, with trailing stems each 1/2 - 5 feet long. Leaves opposite, hairy divided into four to eight pairs of leaflets each about 1/4 - 1/2 inch long and oval. Flowers are yellow, 1/3 - 1/2 inch wide with five petals, borne in the leaf axils. Fruits consist of five sections which, at maturity, break into tack-like structures with sharp, sometimes curving spines, each section two- to four-seeded.

Puncturevine was introduced from southern Europe and is now widely scattered over much of the U.S. It grows in pastures, cultivated fields, waste areas, and along highways and roads. The hard, spiney burs damage wool, are bad in hay and may be injurious to livestock. The seed will remain dormant in the soil for four to five years, which makes eradication difficult. Because of its sharp burs, puncturevine has been spread over a wide area by animals and vehicles. Flowering and seed production occur from July to October.

Seedling puncturevine plants are pinnately compound with hairs appearing on each leaflet.

Puncturevine flowers, yellow in color with five petals, develop into circular spiney fruit that break into individual seeds when ripe.

Velvetleaf
Abutilon theophrasti (Medic.)

Velvetleaf
Malvaceae
(Mallow family)

This is an annual, completely covered with soft hairs, 2 - 7 feet tall; stems erect and branched; leaves alternate, heart-shaped, pointed at the apex, 5 inches or more in width and attached on slender petioles. Flowers are solitary in leaf axils with five yellow petals and numerous fused stamens that form a tube. Seed pods are rounded with nine to 15 arranged in a disc, each containing three to nine egg-shaped, somewhate flattened, rough gray-brown seeds.

Introduced into this country from Asia, velvetleaf is widely distributed in the eastern U.S., but is found only sparingly in Wyoming and Utah. Velvetleaf thrives on rich soils and is found in cultivated fields, gardens, fence rows and waste areas. Velvetleaf seeds retain their viability in soil for more than 50 years, making eradiction very difficult. Flowering and seed production occur from late June to October.

Alternate heart-shaped leaves, up to five inches across, are covered with fine soft hairs.

Velvetleaf seed capsules containing three to nine flattened gray-brown seeds and yellow flowers containing five petals appear in leaf axils from June to October.

Mallow
Malva spp.

Mallow
Malvaceae
(Mallow family)

Mallow is an annual with stems spreading with the branch tips ascending (turning upward), rounded leaves are long-stalked and alternate, with an indention at the base. Leaves range in size from 3/4 to 3 inches in diameter, with five to seven lobes which have rounded teeth. Flowers are stalked, often striped, ranging in color from white to lavender. They contain five petals, which are held by a calyx half the length of the petals. Fruit appears as a button-like structure when immature, but consists of a circle of one-seeded round lobes which separate at maturity. Malva is a genus of about 40 species, several of which are in our area.

Mallow, a plant of European origin, is common in most of the United States. Blooms appear from late June to September. Mallow is not a highly competitive species, but it is common in cropland, waste areas and gardens.

Seedling plants germinating in early spring have round, lobed leaves with a V-notched base.

Lavender-striped flowers and round button-like seed capsules are found in mid-summer on mallow.

Common St. Johnswort
(*Hypericum perforatum* (L.)

Common St. Johnswort
Clusiaceae
(St. Johnswort family)

St. Johnswort is a perennial reproducing by seeds or short runners. Stems are 1 - 3 feet high, erect with numerous branches, somewhat two-ridged, rust-colored, woody at their base. Leaves are opposite, sessile, entire, elliptic to oblong, covered with transparent dots, not over 1 inch long. Flowers are 3/4-inch in diameter, bright yellow, numerous in flat-topped cyme with five separate petals with occasional minute black dots around the edges. Petals are twice as long as the sepals; stamens are numerous, arranged in three groups. Seed pods are 1/4 inch long, rust-brown, three-celled capsule each with numerous seeds.

St. Johnswort, originally from Europe, is frequently found on dry sandy or gravelly soils. The weed contains a toxic substance which affects white-haired animals. Affected animals rarely die, but they will often lose weight and develop a skin irritation when exposed to strong sunlight. St. Johnswort is an abundant weed in Australia. Research conducted there identified three insect species native to Europe which selectively feed on the plant. Insects have been introduced to the Western United States and Canada which, to some extent, has provided a way to control the weed.

Leaves of St. Johnswort are oval in shape with prominent veins. Tiny pin holes are visible when leaves are held up to a light source.

Yellow flowers with five petals and many stamen appear in early summer on St. Johnswort.

Western waterhemlock
Cicuta maculata (L.)

Western waterhemlock

Apiaceae
(Parsley family)

This perennial is native to the intermountain region and is highly poisonous. Erect stems are 3 - 7 feet tall, usually swollen at the base. Leaves alternate, one per node, petioled, pinnately divided. Leaf veins which terminate at the bottom of leaf serrations distinguish this species from others in the family. Flowers are white in compound stemmed umbels mostly flat on top. Each flower is two-seeded. Seeds are somewhat kidney-shaped with corky ridges and tea-colored. A horizontally-divided, enlarged taproot is its most easily-recognized feature. Syrup in the taproot is extremely poisonous to animals and humans.

Western waterhemlock is a wetland plant especially common on pastures or untilled areas. It occurs on streams and irrigation canals. It begins growth in spring, flowers in late spring to early summer. Western waterhemlock, considered one of the most poisonous plants in North America, is often mistaken for edible members of this family. Several deaths each year occur because of this species.

Toothed, pinnately-divided alternate leaves with veins terminating in the bottom of serrations are useful in identifying this extremely poisonous species.

When roots are split, horizontal chambers are found in western waterhemlock. This area is the most poisonous part of the plant. Extreme care should be taken when examining this species.

Poison hemlock
Conium maculatum (L.)

Poison hemlock
Apiaceae
(Parsley family)

Poison hemlock is a tall biennial native to Europe that grows 6 - 8 feet tall with occasional plants growing to 10 feet tall. Stems are erect, stout and purple-spotted with distinct ridges and extensively branched. Leaves have a shiny green color, finely pinnately divided three or four times and leaflets are segmented and 4 - 8 mm long. Lower leaves on long stalks clasp the stem upper leaves on short stalks. Foliage has strong, distinct parsnip odor. Flowers are borne in many umbrella-shaped clusters, each supported by a stalk. Flowers are white with sepals lacking. Seeds are borne in pairs, light brown color about 1/8 inch long. Distinctly ribbed seeds are concave in shape.

It occurs on borders of pastures and cropland, gradually invading perennial crops (i.e. alfalfa). Poison hemlock tolerates poorly-drained soils and frequents stream and ditch banks. All plant parts are poisonous including the large white taproot. Humans have been poisoned by mistaking the plant for parsley.

Stems of poison hemlock are covered with purple spots at all growth stages.

Leaves on mature plants, as well as seedling plants (shown here), are fern-like in appearance because they are divided twice and sometimes three times.

Wild carrot
Daucus carota (L.)

Wild carrot
Apiaceae
(Parsley family)

This biennial herb stands 2 - 4 feet tall. It exists as rosette with a deep taproot the first year. Stems are erect, hollow, stiff-haired and sometimes branched. Leaves alternate, stalked near the stem base. Sessile leaves above. Twice-pinnately compound has narrow segments to 5 inches long. Leaf margins and veins have short hairs. Stong carrot odor. White flowers grow in flat-topped umbels 3 - 6 inches across, five petals. Seeds 1/8 inch long, grayish-brown, one side flattened and rounded side distinctly ribbed. Mature seed with barbed prickles.

Wild carrot is a pernicious weed of older pastures and meadows, strongly discouraged by cultivation. Wild carrot, also know as Queen Anne's lace, has an aromatic odor, and begins growth in early spring, frequents dry areas and rocky soils.

Pinnately-compound leaves grow up to five inches long. Leaf margins and veins are covered with short, stiff hairs.

Wild carrot has flat-topped umbel flowers, similar to poison hemlock and western waterhemlock.

Cow parsnip
Heracleum lanatum (Michx.)

Cow parsnip

Apiaceae
(Parsley family)

This introduced biennial from Europe reproduces by seed, forming a low-growing rosette its first year with large, fleshy taproot. Flower stalks grow 2 - 4 feet tall. Stems are somewhat hairy and grooved. Leaves are glabrous, once pinnately-compound with serrated edges, not carrotlike. Cream colored flowers form with five petals in umbels at the top of short stalks. Flower clusters are mostly flattened with outside flower stalks curving inward with maturity. Seed is flattened on one side, rounded on the other, with distinct ridges.

Cow parsnip occurs mostly in disturbed areas and along roadways. It now inhabits many sites in intermountain regions, frequenting drier sites. Its genus name means parsnip in Latin.

Leaves, commonly 6 - 10 inches across, are deeply divided, not pinnately compound like other parsley family plants that are poisonous.

Cow parsnip has flat-topped umbel flowers which resemble poisonous plants in this family.

Showy milkweed
Asclepias speciosa (Torr.)

Showy Milkweed
Asclepiadaceae
(Milkweed family)

Showy milkweed is perennial, 2 - 5 feet tall, reproducing by seeds and underground rootstocks. Leaves opposite, oval-shaped, prominantly veined, 4 - 7 inches long, and covered with fine, soft hairs. The plant has a greyish-green color and grows erect. All foliage parts exude a milky latex sap when cut. Flowers are arranged in umbels at the top of the plant and are purplish-pink. Reddish-brown flat seeds are borne in 3 - 5 inch spindle shaped pods. Each seed bears a tuft of hairs allowing them to be spread by wind.

Showy milkweed is native to North America and is common along roadsides, ditchbanks, pastures and cultivated fields. Colonies form by spreading rootstocks when the plant is not disturbed by tillage practices.

Seed capsules 3 - 5 inches long burst open in late summer, producing dozens of flat, reddish-brown seeds. Fibers produced within seed capsules are used commercially.

Clustered flowers, purplish-pink in color, appear in mid-summer on showy milkweed.

Field bindweed
Convolvulus arvensis (L.)

Field bindweed
Convolvulaceae
(Morningglory family)

Field bindweed is a perennial from an extensive root system, often forming dense tangled mats. Stems are prostrate, 1 - 4 feet long; leaves alternate, more or less arrowhead. They are shaped with spreading, pointed or blunt holes at the base. The flowers are bell- or trumpet-shaped, white to pinkish, approximately 1 inch in diameter. Fruit is a small, round capsule usually four-seeded. Seeds are dull brown, roughened, flattened on two sides, round on the other and about 1/6 inch long.

Field bindweed was introduced from Europe and has become a widespread and serious weed problem in all parts of the U.S. except the southeastern states. In Wyoming and Utah, it is extensively distributed in cultivated fields and waste places. Because of its remarkable adaptability to different environmental conditions, it may be found at altitudes as high as 10,000 feet. It is a most difficult weed to eradicate because of the large, fleshy, deep taproot which can penetrate the soil to a depth of 10 feet and which repeatedly gives rise to numerous long rhizomes. Seeds remain viable for up to 50 years. The flowering period is from late June until frost in the fall.

Leaves of field bindweed are shaped like arrowheads. They are normally slender with sharp pointed lobes, though many variations occur.

Flowers are funnel-shaped and vary in color from white to pink. Flowering begins in early summer and continues through fall.

Field dodder
Cuscuta spp.

Field dodder

Convolvulaceae
(Morningglory family)

This parasitic plant lacks chlorophyll. Stems are yellowish, thread-like and twining, leaves are reduced to thead-like scales. Flowers are numerous in compact cluster, five parted and shallowly cupped, white to pink. Fruit is a two to four seeded globular capsule which is somewhat depressed on the top. Seeds are small, oval, gray to red and 1/25 inch in length.

Dodder is widely distributed over much of the U.S. with several species present in Wyoming and Utah. Dodder seeds germinate on the soil surface and the resulting plant develops a small root system and two to four inch thread-like stalk which attach to green plants. Once attached, the root system disappears and the dodder becomes wholly parasitic. Many plants serve as hosts for this parasite, but alfalfa and clover are especially susceptible. Dodder seeds are fairly long-lived in the soil and infestation may occur in areas where host plants were not grown for several years. Flowering period is July to October.

Seeds of field dodder germinate in early spring and seedlings quickly attach to host plants to obtain food supplies. Herbicides to control this plant should be applied at germination time.

This parasitic annual attaches to alfalfa and produces many compound flowers in late summer.

Houndstongue
Cynoglossum officinale (L.)

Houndstongue
Boraginaceae
(Borage family)

Houndstongue is a biennial which grows one to four feet tall and reproduces by seed. Leaves lack teeth or lobes, alternate one to 12 inches long, one to three inches wide, rough hairy. Flowers are reddish-purple and terminal. Fruit divides into four prickly nutlets each about 1/3 inch long.

Houndstongue is an introduced plant, native of Europe, that forms a rosette the first year and sends up a flowering stalk the second year. The leaves are rough and resemble a hound's tongue. It may be found in pastures, along roadsides and in disturbed habitats. The nutlets break apart at maturity and cling to passing animals which scatter them. Houndstongue is toxic, containing pyrrolizidine alkaloids, causing liver cells to die slowly. Animals may survive for six months or longer after they have consumed a lethal amount. Sheep are more resistant to houndstongue than are cattle or horses. Horses may be especially affected if confined in a small area with houndstongue and a lack of desirable forage. Therefore, ranges and pastures should be maintained to encourage production of grasses and high quality forage.

The first year of the two year growth cycle of this biennial, rosettes are formed. Herbicides are most effective when applied to this growth stage.

The second year, houndstongue plants flower and produce an abundant supply of seed that is spread by attaching to clothing or animals.

Western sticktight

Lappula occidentalis redowskii (Hornem.) Greene

Western sticktight
Boraginaceae
(Borage family)

This annual grows 6 - 12 inches tall, reproducing by seed. Leaves are narrow, somewhat blunt-tipped. Flowers are numerous, very small, blue. Fruit is divided into four nutlets; nutlet margins covered with a single row of spear-like spines.

Western sticktight is native to the western plains region and is common along roadsides, waste areas, and abused ranges. The spiney seed, becoming entangled in wool, makes it a problem for sheep ranchers. It also readily clings to any passerby, so it is easily distributed.

Seedlings from this annual plant have oblong leaves covered with hair.

Western sticktight produces an abundant supply of seed from mid-summer to fall. Seed is about 1/8 inch in size, spread by clinging to animals and clothing.

Black henbane
Hyoscyamus niger (L.)

Black henbane
Solanaceae
(Nightshade family)

Black henbane may be annual or biennial, 1 - 3 feet tall. Leaves are coarsely-toothed, shallowly lobed, pubescent. Foliage has a foul odor. Flowers from axils of upper leaves are brownish-yellow with a purple center and purple veins. Fruits are approximately 1 inch long, five-lobed.

Black henbane is a native of Europe and has been cultivated as an ornamental. It has spread throughout the United States and is a common weed of pastures, fencerows, roadsides, and waste areas. Black henbane contains tropane alkaloids that have caused occasional livestock poisoning. However, the plant is usually not grazed by animals and is consumed only when more palatable forage is not available. Henbane alkaloids have been used in the past, and are currently used, as medicines at controlled dosages. It is considered a poisonous plant to humans.

Large rosettes having serrated leaves covered with fine hair start growth in early spring. The plant has a pungent odor at all growth stages.

Two rows of pineapple-shaped fruit, about 1 inch long, appear in early fall. Within each capsule are hundreds of tiny black seeds.

Virginia groundcherry
Physalis virginiana (Mill.)

Virginia groundcherry
Solanaceae
(Nightshade family)

This perennial is 12 - 30 inches tall, reproducing by seeds and rootstalks. Stems are hairy, erect, widely branching. Leaves are alternate, oval, 2 - 3 inches long, wavy to bluntly-toothed margins. Flowers are bell-shaped, yellow, with dark centers. Fruit is round, 3/8 - 1/2 inch in diameter, with many seeds. Seeds are 1/17 inch in diameter, yellow, oval, flattened.

Virginia groundcherry is a native plant in the same family as potatoes and tomatoes. It grows in cultivated fields, along roadsides, ditches and in waste areas. It is especially troublesome in bean fields. Control of this perennial is difficult. Identification is difficult because of the similarity of species.

Immature plants of Virginia groundcherry have narrow leaves, 2 - 3 inches long with bluntly toothed margins.

Bell-shaped, yellow flowers with dark centers develop into a bladder-like, pear-shaped case which encloses the seed-bearing fruit.

Silverleaf nightshade
Solanum elaeagnifolium (Cav.)

Silverleaf nightshade
Solanaceae
(Nightshade family)

Perennial, 12-36 inches tall with creeping rhizomes and scruffy pubescence that gives the plant a silvery appearance. Leaves are lanceolate with wavy margins, while plant stems have yellow thorns. Flowers are 3/4 - 1 inch broad with violet to bluish petals and prominent yellow stamens. The mature fruit is a dull orange to yellow berry.

Silverleaf nightshade is native to the central United States; but it has spread to other areas where it is found on wasteland, pastures and cropland. The berries and foliage are poisonous to livestock. It is resistant to several commonly used herbicides.

Stems are covered with scruffy, short hairs and have yellow thorns.

Flowers of silverleaf nightshade are blue to violet in color and have five petals with yellow centers. Fruit are yellow or orange and are filled with seed.

Black nightshade
Solanum nigrum (L.)

Black nightshade
Solanaceae
(Nightshade family)

This weed is an annual, 6-24 inches tall with glabrous or appressed-hairy stems. Leaves are ovate, wavy-edged and tapered to the tip. Flowers resemble those of the potato and tomato and are white to pale-blue, 1/4 - 3/8 inch wide, borne in clusters. Calyx is scarcely accrescent, not cupping the black fruit.

Black nightshade, a native of Europe, is a weed of waste places and cultivated fields. The berries frequently become mixed with harvested commodities such as dry beans and decrease crop quality. The green (immature) fruit and the foliage contain toxic alkaloids. In recent years, nightshades have become troublesome in fields where certain herbicides have been used that control other weeds but are weak on nightshade.

Black nightshade seedling leaves are ovate and taper to a pointed tip.

Fruit of black nightshade are black when mature. The calyx, or covering on the top of the berry, is smaller than on hairy nightshade. Stems and leaves have a smooth appearance.

Buffalobur
Solanum rostratum (Dunal)

Buffalobur
Solanaceae
(Nightshade family)

This native annual grows to 2 feet high when mature, has leaves 2 - 5 inches long with deep lobes and is covered with spines. Yellow, five-lobed flowers common throughout the summer measure about 1 inch across. The berry is enclosed by an enlarged calyx and contains many seeds. Seeds are black, wrinkled and flattened.

Buffalobur is widely distributed in the western U.S. It is common on disturbed wastelands and prairies. It is most common on sandy soils, but grows on most soils. It is very drought resistant and serves as a host for the Colorado potato beetle. It is not a highly competitive species and can be controlled with herbidices.

Seedling plants of buffalobur are deeply lobed and spiney with prominant veins.

Yellow flowers appear in early summer and develop into a spiney seed capsule containing many small seeds.

Hairy nightshade
Solanum sarrachoides (Sendt.)

Hairy Nightshade
Solanaceae
(Nightshade family)

Hairy nightshade is an annual, 12-24 inches tall with spreading hairy pubescence on the foliage. Foliage may feel sticky when handled. Flowers resemble those of the potato and tomato and have five white petals, and an accrescent green calyx. They are arranged in clusters. As the fuit matures, it cups the lower half of the fruit. The fruit is greenish or yellowish.

Hairy nightshade, a native of South America, is a widespread weed of waste places and clutivated fields. The plant contains toxic alkaloids, especially in the berries. Hairy nightshade causes problems in field crops similar to those described for black nightshade.

Immature plants start their growth in early summer. Leaves and stems are quite hairy.

White flowers having five petals with yellow centers develop into green fruit containing very small seeds. Each berry is half-covered with a calyx.

Cutleaf nightshade
Solanum triflorum (Nutt.)

Cutleaf nightshade

Solanaceae
(Nightshade family)

This plant is an annual, 4 - 24 inches tall, branched from the base, with ill-smelling foliage. Leaves are deeply lobed and may have short pubescence. Flowers resemble those of the potato and tomato and are small, have white petals and a somewhat accrescent calyx and are arranged in groups of three. Fruits are green.

Cutleaf nightshade is native to North America east of the Cascade Mountains and is a weed of waste places and cultivated fields. The plant contains toxic alkaloids. Cutleaf nighshade is resistant to many commonly used herbicides.

Seedlings of cutleaf nightshade have deeply lobed leaves and are covered with short hair. Seedlings commonly appear in early spring.

Flowers of cutleaf nightshade are white with five petals and yellow stamen centers. Fruit are abundant and contain many yellow seeds.

Dalmatian toadflax
Linaria dalmatica (L.) Miller

Dalmatian toadflax
Scrophulariaceae
(Figwort family)

Dalmatian toadflax is a perennial, grows up to 3 feet tall, reproducing by seed and underground root stalks. Leaves are dense, alternate, entire, upper leaves are conspicuously broad-based. Flowers are borne in axils of upper leaves and bracts, flowers two-lipped, 3/4 - 1 1/2 inch long with a long spur, yellow with an orange, bearded throat, fruit a two-celled capsule with many irregularly angled seeds.

Dalmatian toadflax was introduced from southeastern Europe and was probably introduced as an ornamental to the United States. It is aggressive and may be found along roadsides and in rangeland where it may become a serious problem by crowding out desirable forage. An extensive and deep root system along with a waxy leaf make this an extremely difficult plant to control.

Early spring growth of this prolific perennial has waxy leaves with a blue-green color. Leaves individually clasp the stem.

Yellow flowers with long spurs appear at mid-summer with plants continuing to flower until early fall.

Yellow toadflax
Linaria vulgaris (Mill.)

Yellow toadflax
Scrophulariaceae
(Figwort family)

Yellow toadflax is perennial, 1 - 2 feet tall; reproducing by seed and underground rootstocks. Leaves are pale green, numerous, narrow, pointed at both ends, 2 1/2 or more inches long. Flowers are 1 inch long with a bearded, orange throat. Fruit is round, 1/4 inch in diameter, brown, two-celled with many seeds. Seeds are dark brown to black, 1/12 inch in diameter, flattened with a papery circular wing.

Yellow toadflax is a native of Eurasia, introduced to the United States in the mid-1800s as an ornamental. This creeping perennial is an aggressive invader of rangelands displacing desirable grasses. It is also found along roadsides, waste places, and cultivated fields. An extensive root system makes this plant difficult to control.

Narrow leaves which are pointed at both ends are less than 2 1/2 inches long and are individually connected to the central stem.

Yellow flowers with an orange throat have spur-like appendages.

Common mullein
Verbascum thapsus (L.)

Common mullein

Scrophulariaceae
(Figwort family)

This biennial produces a large thick rosette of fuzzy leaves the first year and a single, stout, erect stem 2 - 6 feet tall the second year. The leaves are alternate, overlapping one another, light green, densely woolly. Flowers are sessile, borne in long terminal spikes, sulfur yellow, five-lobed and more than an inch in diameter. Fruits are two-chambered with numerous, small, angular, brownish seeds, 1/32 inch long.

This weed was introduced from Europe, but it is a native of Asia and is quite common throughout the temperate parts of North America. It is a common sight along river bottoms, in pastures, meadows, fence rows and waste areas especially on gravelly soils throughout Wyoming and Utah. Because of the large number of seeds produced by each plant, it is difficult to control. Livestock will not eat the plant because of its wooliness. Flowering and seed production occur from June to August.

Common mullein produces a rosette the first year of its two-year life cycle. Leaves of this plant are covered with soft, fine hair.

Yellow five-lobed flowers are borne on long seed stalks in late summer.

Buckhorn plantain
Plantago lanceolata (L.)

Buckhorn plantain

Plantaginaceae
(Plantain family)

This is a perennial from a stout rootstock and fibrous roots. The crown of the plant is covered with tan woolly hairs. Leaves all basal, three to several veined, narrlowly elliptic or lance-elliptic, 4 - 12 inches long, usually less than 1 1/2 inches wide. Flowering stalks are several, up to 18 inches tall, bearing a dense spike of small flowers, subtended by thin ovate bracts; stamens-four, exerted, conspicuous; four shiny seeds per fruit, blackish, about 1/16 of an inch long. *Plantago* is a genus of more than 200 species, two of which are weedy in our area.

Buckhorn plantain, also called English plantain, is native of Eurasia and is a weed of roadsides, pastures and other disturbed sites such as lawns and gardens. Flowering period is May to August.

Seedlings of buckhorn plantain have long narrow leaves with prominent parallel veins.

Seed stalks are produced in late summer. White flowers surround the base of the seed stalk.

Broadleaf plantain
Plantago major (L.)

Broadleaf plantain
Plantaginaceae
(Plantain family)

This is a fibrous-rooted perennial, sometimes blooming the first year, from an erect caudex, crown not woolly. Leaves are all basal, broadly elliptic to ovate, blade abruptly contracted to the well-defined petiole. The blade is entirely or irregularly toothed, 3 - 7 inches long, three to several-veined. The flowering stem is 5 - 15 inches long, topped with a dense, elongated spike of small flowers. Corolla lobes are papery, reflexed; stamens-four, exerted. Fruit is 1/4 inch long; seeds 6 - 30, black or brown, strongly reticulate.

Broadleaf plantain is supposedly a native of Europe; but today it has a cosmopolitan distribution, inhabiting roadsides, lawns, disturbed sites and cultivated fields. It can be found in valleys to midmontane sites and flowers from April to September.

In early spring, this perennial forms a rosette which has leaves from 3 - 7 inches long with wavy margins and prominant veins.

Seeds are borne on terminal stems without pedicles or short stems attaching them to the main stem.

Common teasel
Dipsacus sylvestris (Huds.)

Common teasel
Dipsacaceae
(Teasel family)

Teasel is a stout, taprooted biennial which grows to 6 feet tall, branched above. The stem is striate-angled with several rows of downward turned prickles. Leaves are conspicuously veined, with stiff prickles on the lower midrib. The basal rosette of leaves usually dies early in the second season. Stem leaves lanceolate up to 10 inches long, the opposite leaves of the stem have fused bases which trap rain water. Flowers are purple, borne in dense heads, each flower subtended by spine-like bractlets. Involucral bracts at the base of the head are generally longer than the head; corolla is four-lobed. Fruits are four-angled, each with a single seed.

Common teasel is a native of Europe, now widespread as a weed in North America. It is found commonly in Utah, in moist sites, especially along irrigation ditches, canals and disturbed sites. While some do not consider this weed important, it is spreading rapidly throughout the Rocky Mountain states. In the fall the stems and fruiting heads are commonly silvered or gilded for winter bouquets. Flowering occurs from July to August.

Teasel is a biennial plant producing a rosette the first year. It has a wrinkled appearance and spines on the midrib of the leaf.

In the second year of the growth cycle, teasel produces spiney seed heads often reaching lengths over 2 inches. Purple flowers are protected by spine-like bractlets.

Common yarrow
Achillea millefolium (L.)

Common yarrow
Asteraceae
(Sunflower family)

This is an aromatic perennial with stems arising singly or loosely clustered form a weakly spreading rhizome, 1 - 2 feet tall, with woolly hairs. Leaves equally distributed along the stem; blade lanceolate in outline but finely dissected into many ultimate segments, overall dimensions are 2 - 6 inches long and 1/4 - 1 inch wide. Inflorescence with many small heads in a flat-topped or dome-shaped cluster. It will have about five ray flowers, white or pink, and 10 - 20 disk flowers.

This species includes a complex of several, ill-defined phases and polyploid races. It is widely distributed throughout the temperate northern hemisphere. Yarrow is reluctantly included here as a weed which is widespread occupying many plant communities including aspen, conifer, sagebrush, mountain brush, riparian, meadow and alpine at 4,000 - 11,000 feet. The flowering period is June to August. Its elimination should only be on private land after careful evaluation.

Leaves in early spring are fern-like looking similar to soft blue-green feathers.

Flowers are borne on terminal stems giving plants a flat-topped appearance.

Common ragweed
Ambrosia artemisiifolia (L.)

Common Ragweed

Asteraceae
(Sunflower family)

This native annual reaches four feet tall reproducing by seeds. Stems and leaves are blue-green in color and are covered with fine hairs. Leaves are pinnately divided and are both alternate and opposite. The bottom side of the leaf has a gray appearance because of fine hairs. Flowers are found on terminal branches, which have male and female flowers in clusters with staminate flowers found above the pistillate flowers. Seeds are awl-shaped with a spiney projection on one side. Flowering occurs in late summer, with seed maturity completed by October.

This annual is common throughout the western U.S. causing hay fever for many individuals. Even though it is common, it is not highly competitive in crops or rangeland. It is commonly found along ditches and in waste areas.

Seedlings appear in early spring for this annual weed. Leaves are deeply-lobed with rounded tips.

Numerous flowers appear in early fall on terminal stems. Pollen is produced at that time, causing many people to have hay fever.

Skeletonleaf bursage
Ambrosia tomentosa (Nutt.)

Skeletonleaf bursage
Asteraceae
(Sunflower family)

Skeletonleaf bursage is a perennial with extensive creeping rootstocks, 4 - 18 inches tall. Stems are branched and somewhat bushy; leaves alternate, with the lower surface covered with minute white hairs. The upper surface is smooth, green, 2 - 5 inches long, ovate to lanceolate in outline, coarsely toothed and deeply segmented. Flowers are inconspicuous and green. Staminate flowers are found in solitary, elongated, terminal clusters; pistillate flowers form in pairs in leaf axils below. Fruit is a light brown bur with conical spines containing one or more achenes.

This weed is a native of the plains region extending from the Missouri River westward. It grows in cultivated fields, pastures, prairies and waste areas. It also survives well under varied soil moisture conditions. Skeletonleaf bursage is a difficult weed to eradicate because of its extensive horizontal root system. Flowering and seed production occur from late June to August.

Leaves are deeply lobed with a dark green upper surface and a silvery-green lower surface which is covered with fine hairs.

Terminal flower clusters with inconspicuous yellow flowers appear in late summer.

Giant ragweed
Ambrosia trifida (L.)

Giant ragweed
Asteraceae
(Sunflower family)

Giant ragweed, a native annual exceeding 10 feet in moist locations, can mature as smaller plants and form seed in drier areas. Stems and leaves are rough, leaves are palmately divided, normally having three lobes. It can also have up to five lobes or be unlobed. Leaves are opposite with long petioles attached to a single center stem. Upper flowers, male (staminate) and lower flowers, female (pistillate) are located in terminal clusters. Seed are over 1/4 inch in length with four or five terminal spikes on each.

Giant ragweed is common in the western U.S. often found in moist areas, near roadsides and wasteland. A major cause of hay fever in the United States. It is not a highly competitive species in crop or rangeland and can be controlled at early growth stages with herbicides.

Large, deeply-parted leaves are normally lobed three times, but may contain up to five lobes.

Flower clusters, often over 6 inches long, appear in the fall. Pollen shed at that time causes people to have hay fever problems.

Common burdock
Arctium minus (Hill) Bernh.

Common burdock
Asteraceae
(Sunflower family)

Common burdock is a biennial, producing a rosette of large, cordate, thickly hairy leaves the first year and an erect, much branched, coarse stem 3 - 10 feet tall the second year. The leaves are alternate, large, broadest at the base, somewhat diminished upwards, margins toothed or wavy, woolly beneath at least when young, dark green above. Flowers are purple, heads borne in leaf axils or at the end of branches, numerous, clustered, covered with many slender (often hooked) spines, achenes gray to brown, mottled, oblong, about 1/4 inch long, flattened and slightly curved.

Native of Europe, common burdock is now established throughout much of the U.S. In Wyoming and Utah, it is found growing mainly along roadsides, ditchbanks, in pastures and waste areas. The burs can become entangled in the hair of livestock allowing seed to be distributed to new areas. Flowering and seed production occur from July to September.

The first year common burdock produces large leaves with no flowers. At this growth stage burdock can be controlled most effectively with herbicides.

Purple flowers are borne in this bur-like structure from mid-summer until fall. Burs hooked on the end gave rise to the idea for Velcro.

Common sagewort
Artemisia campestris (L.)

Common sagewort
Asteraceae
(Sunflower family)

Common sagewort is a biennial forming a rosette the first year then a seed head the second. Leaves are long and sometimes divided into three segments that are thread-like, with long petioles. Plants produce seed heads up to 3 feet long. Upright stems are covered with leafy branches on their upper half. Flowers located on short stalks are erect nodding when mature in August and September. Seed maturity occurs in October.

This biennial plant is common throughout pastures and rangeland and is often found along ditches and near roads. It is not considered a highly competitive plant. Herbicides are used when areas become overpopulated with this species.

The first year's growth of common sagewort of its two year (biennial) life cycle is blue-green with leaves that are deeply lobed.

Flowers borne along stems mature in August and September.

Silver sagebrush
Artemisia cana (Pursh)

Silver sagebrush

Asteraceae
(Sunflower family)

Silver sagebrush is a perennial capable of reproducing by seeds and resprouting from roots when topgrowth is destroyed. Leaves are simple, alternate and lance shaped, occasionally leaves have one or two irregular teeth. Leaves are covered with fine hair giving them a silver color. Flowers appear in August and September within leafy panicles. Seed capsules are round, have four or five ribs and are shed in October to December. Yellowish stems reach heights of over 6 feet when plants mature.

Plants become highly competitive with perennial grasses. They generally increase with cattle usage and decrease in population with sheep usage. Plants are used as browse for wildlife and sheep. Habitats include loamy and sandy soils. Control measures such as herbicides and fire are only partially effective because of the resprouting nature of this woody species.

Leaves, covered with hair, are simple, alternate and lance-shaped ending in a single point.

Yellow flowers form in late summer along the stems of silver sagebrush.

Sand sagebrush
Artemisia filifolia (Torr.)

Sand sagebrush

Asteraceae
(Sunflower family)

Sand sagebrush, a perennial, grows to heights of 3 feet and reproduces from seed. Stems are fine in nature, woody and branched, having a gray appearance from pubescence on leaves and stems. Leaves are threadlike from 1 1/2 - 3 inches in length. Fruit are borne in upper leaf axils as a rounded structure. Flowers form a plume-like structure from July to October, with seed maturity taking place in October.

Sand sagebrush grows on sandy, well-drained soils in rangeland and pastures. It is highly competitive with perennial grasses and can be controlled with herbicides.

Leaves of this perennial woody, species are thread-like, growing from 1 1/2 - 3 inches in length.

Seedling plants of sand sagebrush have thread-like leaves with fine, woody stems.

Fringed sagebrush
Artemisia frigida (Willd.)

Fringed sagebrush
Asteraceae
(Sunflower family)

Fringed sagebrush is a perennial reproducing by seeds. Plants reach heights of 1 1/2 feet when groups of mature seed heads are formed. Plants are mat-like with a silvery color before producing flower stalks. Seeds are borne on single seed stalks which appear as rounded structures. The plant has a sage odor, is silvery blue in color and is bitter to taste. Leaves are finely divided into either three or five segments up to 1 1/2 inches long. Flowers are yellow, discoid in shape and are borne within long hairy bracts having a panicle shape. Achene or oblong seed are flattened with rounded edges and are gray to brown in color. Flowering begins in August with seeds maturing in September.

Fringed sagebrush is an important wildlife and sheep feed source, but is highly competitive with perennial grasses needed by cattle. It is widespread on rangeland and increases rapidly with overgrazing. Herbicides are effective in controlling this species.

Early spring growth of fringed sagebrush appears as blue-green clusters within a gray background.

Fringed sagebrush, a mat-forming plant, produces seed stalks in late summer and fall with round yellow flowers among hairy bracts.

Big sagebrush
Artemisia tridentata (Nutt.)

Big sagebrush
Asteraceae
(Sunflower family)

Big sagebrush is a perennial reproducing from seed. Plants average less than 3 feet in height; but in deep soils this woody species reaches heights over 10 feet. It is considered an evergreen even though leaves have a grayer color in winter months. Leaves are undivided, have a wedge shape and three blunt lobes. Leaves contain essential oils causing them to have a distinct sage odor when they are crushed. They are silvery blue in color because of the dense gray hairs found on both sides. Yellow flowers appear in late August, forming panicles. Seed formation and shed takes place from October to December. Seed have a one-year viability once they are shed. Older plants have trunks over three inches, which are covered with brown stringy bark.

Big sagebrush is a highly competitive species and often limits perennial grass production to 1/3 that of areas which have been cleared of sagebrush. It is the most common woody species in western U.S. rangelands. Control cannot be obtained by grazing management alone. Excellent control has been attained for over 20 years by treating with herbicides, burning or mechanical removal of the species.

This woody perennial has blue-green strap-shaped leaves which have three distinct lobes at their tips.

Flowers are formed in August with seed being shed from September to December.

Plumeless thistle
Carduus acanthoides (L.)

Plumeless thistle
Asteraceae
(Sunflower family)

Plumeless thistle is a winter annual or biennial herb, rarely flowering the first year, from stout fleshy taproots. Stems grow from 1 - 4 feet tall; they are freely branched above and covered with spiny wings 1/4 to 3/4 inch wide extending up to the flowering heads. Basal rosette leaves are usually 4 - 8 inches long with spinose lobes. Stem leaves alternate, sessile and blending into the stem. Peduncles usually spiny-winged up to the base of the flowering heads; heads solitary at the ends of branches or in clusters of two to five. Involucral bracts narrowly lanceolate, sparsely to densely hairy. Corolla mostly purple, rarely white or yellowish.

This native of Eurasia is infrequent to locally abundant in pastures, stream valleys, fields and roadsides. While it isn't known in Utah, it is frequent in Idaho, eastern Wyoming and has the potential of becoming a noxious weed component in Utah. Blooms occur from May to July.

A seedling rosette of plumeless thistle has wavy leaves with yellow spines along the leaf margins.

Flowers, 1 - 2 inches in diameter, are purplish-pink with bracts holding the seed appearing as sharp spines. Flowering occurss from May to July.

Musk thistle
Carduus nutans (L.)

Musk thistle
Asteraceae
(Sunflower family)

Musk thistle is biennial or sometimes a winter annual, which grows up to six feet tall. Leaves are dark green with light green midrib, deeply lobed, spiney margined. These leaves extend onto the stem giving a winged appearance. Flower is terminal, solitary, 1 1/2 - 3 inches in diameter. It is usually bent over. Flowers are deep rose, violet or purple, occasionally white; they are subtended by long, narrow, spine-tipped bracts. Fruits are 3/16 inch long, shiny, yellowish-brown with a hair-like plume.

Musk thistle was introduced to the U.S. in the early part of the century and is now widespread throughout the U.S. and Canada. It is native to southern Europe and western Asia. It invades pasture, range and forest lands along with roadsides, waste areas, ditch banks, stream banks and grain fields. Its aggressive nature allows it to spread rapidly forming extremely dense stands which crowd out desirable forages. Chemical control is effective. An introduced biological control agent, the musk thistle weevil, feeds on the seeds and will limit the spread of this plant.

Musk thistle seedlings have wavy margins and are somewhat lobed with marginal yellow spines.

Flowers appear on this biennial in mid-summer. Flowers are a deep rose color and measure up to three inches across. Bracts under the flower are a very distinctive identification feature.

Purple starthistle
Centaurea calcitrapa (L.)

Purple starthistle
Asteraceae
(Sunflower family)

Purple starthistle is annual or biennial. Plants are prolific seed producers reaching heights of over six feet when mature. Plants form a rosette in late May and June and flower in July and August. The stems and leaves are covered with fine hairs with leaves being divided into narrow linear segments with the leaf tip narrow and undivided. The bracts of the purple flowers have straw-colored, spinelike projections over one inch in length. Seeds are pale in color and contain no pappus.

Introduced from Europe, this species was reported abundant in California in the early 1950s in some counties. It is found in Converse County, Wyoming, and is contained within a small area. If this species is found it should be treated with a herbicide to keep it from spreading. Inform your County Extension Service or Weed Control Districts of the location of this species.

Immature plants appear in early spring of this biennial. Leaves are deeply lobed with light colored midribs.

Flowers are light purple with yellow spines, over an inch long, extending from the involucre or seed covering of the plant.

Diffuse knapweed
Centaurea diffusa (Lam.)

Diffuse knapweed

Asteraceae
(Sunflower family)

This weed is a diffusely branched annual or perennial, 1 - 2 feet tall, stems are rough to the touch. Leaves are small, pinnately divided; lower leaves are deciduous, the reduced leaves of the inflorescence are mostly entire. Flowering heads are numerous and narrow. Ray flowers are white to rose or sometimes purplish; involucral bracts are divided like the teeth of a comb, tipped with a definite slender spine; pappus is lacking. Achenes are brown or grayish.

Centaurea is a large genus of over 400 species, centering in the Mediterranean region. All of the species treated here have been introduced from Eurasia within the last 100 years and now represent a threat to rangelands. Diffuse knapweed infests roadsides, waste areas and dry rangelands, and as a highly competitive plant, threatens to exclude many desirable forage species. Flowering occurs from July to September.

Seedling stages of diffuse knapweed have finely divided leaves covered with short hair. Herbicides are most effective when applied at this early growth stage.

Bracts under the white flower are yellow spines with teeth appearing as a comb along the spine margins.

Spotted knapweed
Centaurea maculosa (Lam.)

Spotted Knapweed

Asteraceae
(Sunflower family)

Spotted knapweed is a biennial or short-lived perennial with a stout taproot. It can have one or more stems, branched 1 - 3 feet tall. Basal leaves up to six inches long, blades narrowly elliptic to oblanceolate, entire to pinnately parted; principal stem leaves pinnately divided. Flowering head is solitary, with terminating branches; involucral bracts stiff and tipped with a dark comblike fringe. The ray flowers are pinkish-purple or rarely cream-colored. Fruits are about 1/8 inch long, tipped with a tuft of persistent bristles.

Spotted knapweed, which was introduced from Eurasia as a contaminant of alfalfa and clover seed, ranks as the number one weed problem on rangeland in western Montana. Utah and Wyoming can expect a reduction in desirable diverse plant communities if this species is allowed to spread. Knapweeds readily establish themselves on any disturbed soil, and their early spring growth makes them very competitive for soil moisture and nutrients. There is some evidence that knapweeds release chemical substances which inhibit surrounding vegetation. The flowering period extends from June to October.

Spotted knapweed rosettes appear in early spring as deeply lobed leaflets radiating from a common point. Herbicides are most effective when applied at this growth stage.

Flowers of spotted knapweed are pinkish-purple. Bracts under the flowers have dark spots tipped with fringe. Leaves of the mature plant are finely divided.

Russian knapweed
Centaurea repens (L.)

Russian knapweed

Asteraceae
(Sunflower family)

Russian knapweed is perennial, forming dense colonies by adventitious shoots from widely spreading black roots. Stems are erect, openly branched, 18 - 36 inches tall. Lower leaves are deeply lobed, 2 - 4 inches long; upper leaves narrow to a sessile base, entire or serrate. Cone-shaped flowering heads grow 1/4 - 1/2 inch in diameter, solitary at the tip of leafy branchlets. Ray flowers are pink to lavender. Many pearly involucral bracts form with rounded or acute papery tips. Seedlike achenes are 1/8 - 1/4 inch long, bearing numerous whitish bristles.

Russian knapweed is a noxious native of Eurasia, probably introduced in North America about 1898. It is now widely established in the western U.S. This species forms colonies in cultivated fields, orchards, pastures and roadsides. Two characteristics distinguish Russian knapweed from spotted knapweed. First, the involucral bracts of Russian knapweed are egg-shaped with papery tips which may be entire or finely hairy. Second, the Russian knapweed plants spread by black, deep growing roots which penetrate to a depth of over 8 feet. Flowering occurs from June to September.

Leaves of newly emerging plants are toothed and covered with fine hair, giving them a blue-green color.

Flowers of this perennial are pinkish-purple. Bracts under the flower are rounded with pointed papery tips.

Yellow starthistle
Centaurea solstitialis (L.)

Yellow starthistle
Asteraceae
(Sunflower family)

Yellow starthistle is an annual introduced from Europe, found commonly in California, Oregon, Washington and Idaho. It usually grows to a height of 30 inches with rigid branching stems covered with a cottonly fiber. Basal leaves are deeply lobed while upper leaves are entire and sharply pointed. Stems have winged structures. Yellow flowers are located singly on ends of branches, bracts of flowers are armed with sharp straw-colored thorns up to 3/4 inch long. Fruit from marginal flowers are dark-colored without bristles, while inner fruits are lighter in color having a terminal tuft of white bristles.

Yellow starthistle grows on various soil types and is usually introduced on roadsides and waste areas. Rangeland is not utilized in areas where this spiney weed is found. When horses are forced to eat this plant, a nervous disorder known as "chewing disease" results causing them to die from not being able to eat. Infestations of this species can be controlled with herbicides and should be reported to your county extension agent or weed and pest supervisor.

Seedling plants appear in early spring. Leaves have deeply lobed margins with pointed tips.

Yellow flowers appear on this annual plant in mid-summer. Yellow spines up to 3/4 inch long extend from the involucre or seed case.

Oxeye daisy
Chrysanthemum leucanthemum (L.)

Oxeye daisy
Asteraceae
(Sunflower family)

Oxeye daisy is an erect rhizomatous perennial, 10 - 24 inches tall, glabrous to sparsely hairy. Leaves progressively reduce upward. Basal and lower stem leaves are oblanceolate to narrowly obovate, 2 - 5 inches long including the petiole, margin crenate to lobed or parted. Upper leaves become sessile and merely toothed. Flowering heads are solitary at the ends of branches, the disk is 1/2 - 3/4 inches across. Ray flowers are white and 1/2 - 3/4 inches long. Fruits are round with about 10 ribs, with no pappus. This genus of about 100 species is native chiefly to the northern hemisphere, especially in the Old World.

Oxeye daisy is native of Eurasia which has escaped cultivation. Now it can be found in meadows, roadsides and waste places. Flowering occurs from June through August.

Leaves of this creeping perennial have lobed margins with basal leaves growing up to 5 inches long.

Flowers having white ray flowers and yellow disk flowers are often transplanted as ornamentals.

Gray rabbitbrush
Chrysothamnus nauseosus (Pall. ex Pursh) Britt.

Gray rabbitbrush

Asteraceae
(Sunflower family)

This perennial is strongly scented with many branches. Stems grow to 2 - 4 feet tall with a white-woolly or white-silky covering. Leaves are numerous, slender, alternate, not twisted, 3/4 - 2 inches long, white-woolly or white-silky. Flowers are yellow, terminal in large clusters. Fruit is five-angled and hairy.

Gray rabbitbrush also commonly called rubber rabbitbrush is one of at least 70 native species of rabbitbrush that occur mainly in the western U.S. They grow well on dry soils and are often associated with sagebrush. Rabbitbrush is an invader and its presence in dense stands indicates overgrazing. Plants rate low in palatability, however, sheep and wildlife will eat it. This plant was used by Indians as chewing gum, for yellow dye, tea, and for coughs and chest pains. Control of this undesirable shrub is difficult due to its ability to resprout from the crown after being sprayed or burned.

Stems of gray rabbitbrush are covered with silky hair, giving them a white color. Leaves are narrow and straight.

Yellow flowers appear in late summer on stem terminals as large clusters.

Douglas rabbitbrush
Chrysothamnus viscidiflorus (Hook.) Nutt.

Douglas rabbitbrush

Asteraceae
(Sunflower family)

Douglas rabbitbrush is a perennial with stems 10 - 24 inches tall. Its covering is brown. Leaves are narrow, usually twisted, green, smooth or short-hairy but never woolly or silky. A sticky secretion is usually found at the base of the leaves and branches. Flowers are yellow, small, in rounded or flat-topped clusters.

Douglas rabbitbrush is a native shrub similar to gray rabbitbrush in the habitat in which it is found. It is often confused with broom snakeweed. However, Douglas rabbitbrush usually has twisted leaves whereas those of broom snakeweed are straight. In addition, broom snakeweed dies completely back to the ground each year, while Douglas rabbitbrush does not. Like gray rabbitbrush it is considered an undesirable invader, which increases with overgrazing and is difficult to control.

Leaves attach to brown-barked stems, having a spiral appearance. New growth comes from woody stems rather than the crown of the plant.

Flowers are yellow in color and appear as clusters in late summer and fall.

Chicory
Cichorium intybus (L.)

Chicory
Asteraceae
(Sunflower family)

Chicory is perennial from a deep taproot, a plant with milky juice exuding from cut stems and leaves. Stems run 1 - 6 feet tall with spreading branches. Leaves are rough, basal leaves in rosette, oblanceolate, petiolate, toothed or often pinnately parted, and 2 - 10 inches long. Upper leaves become reduced, sessile, and even entire. Flowering heads are borne 1 - 3 together in the axils of much reduced upper leaves. Ray flowers will be blue or occasionally white, sometimes with a spread of 1 1/2 inches. Fruits are weakly angled or ribbed, tipped by a crown of minute scales.

Chicory is a native of the Mediterranean region, but now it is cosmopolitan in the northern hemisphere. It is widespread along roadsides and disturbed sites, having been planted by man for use as salad greens and the root as a substitute for coffee. Flowering occurs from July to September.

Immature growth of chicory appears as a rosette with lobed leaves that are rough in appearance.

Flowers up to 1 1/2 inches across are normally blue or purple, sometimes white, and appear in mid-summer. Flower petals are square on the end and lobed.

Canada thistle
Cirsium arvense (L.) Scop.

Canada thistle
Asteraceae
(Sunflower family)

Canada thistle is a colony-forming perennial from deep underground and extensive horizontal roots. Stems are 1 - 4 feet tall, ridged, branching above. Leaves are alternate, lacking petioles, oblong or lance-shaped, divided into spiny-tipped irregular lobes. Flowers are unisexual, on separate plants; flowers purple in heads 1/2 - 3/4 inch in diameter; involucral bracts spineless. Fruits are about 1/8 inch long, somewhat flattened, brownish, with a tuft of hairs at the top.

Canada thistle is a native of southeastern Eurasia. It was introduced to Canada as a contaminant of crop seed as early as the late 18th century. Canada thistle differs from other species of the true thistle in that there are male and female flower heads, and these are on separate plants. By asexual reproduction, it is possible that a colony of male plants would produce no fruits, but still maintain itself. This aggressive weed is difficult to control; for example, breaking up the roots by plowing only serves to increase the number of plants. Flowering occurs during July and August.

Early spring growth appears as rosettes with spiney-tipped, wavy leaves.

Clusters of purple flowers up to 3/4 inch across appear in late summer on this perennial. Bracts under the flowers are spineless.

Platte thistle
Cirsium canescens (Nutt.)

Platte thistle

Asteraceae

(Sunflower family)

Platte thistle is a perennial reproducing from seed. Seedling plant leaves become lobed as they develop. The entire plant is covered with fine hair. Leaves join the stem, becoming a part of it. Each leaf is deeply lobed with yellow spines at the tip of each lobe. Leaf margins have smaller spines. Flower buds appear in the center of each stem and in branch axils. Flowers are cream-white with the bracts being covered by short yellow spines. Seeds are light colored and curved. Blooms appear in early summer with seed development completed by July.

Platte thistle is very common on roadsides and in rangeland but is not considered a threat to livestock forage production.

Early spring growth appears as a blue-green rosette with deeply lobed leaves.

White flowers appear on this perennial in mid-summer. Yellow spines are found on the rounded involucre under the flower.

Bull thistle
Cirsium vulgare (Savi) Tenore

Bull thistle
Asteraceae
(Sunflower family)

Bull thistle is a biennial with a short, fleshy taproot. The stem is 2 - 5 feet tall, bearing many spreading branches. It is green or brownish, sparsely hairy, irregularly and spiny winged. Leaves in the first year form a rosette, stem leaves are green, glabrous, bearing fine, short, yellowish, appressed prickles, leaves pinnately lobed, the larger ones with the lobes again toothed or lobed. Flowering heads are 1 1/2 - 2 inches wide, more or less clustered at the ends of branches. Involucral bracts are narrow, spine-tipped, progressively longer and narrower from outer to inner ones, flowers dark purple. Young fruits are topped by a circle of plume-like white hairs.

Bull thistle is native of Eurasia and is now widely established in North America, having been introduced many times as a seed contaminant. Pastures, fields, roadsides and disturbed sites are all potential habitats for this highly competitive weed. Flowering occurs from July through September. It is possible to separate bull thistle from Canada thistle by examination of the leaves alone. Bull thistle leaves are prickly hairy above and cottony below, while the Canada thistle leaves are glabrous above and glabrous or hairy below.

A rosette appears the first year of this biennial's life cycle. Leaves have prickles on the surface with margins deeply lobed and wavy.

Flowers of this biennial are pinkish-purple and appear in mid-summer. The involucre is somewhat tapered and covered with spines.

Rush skeletonweed
Chondrilla juncea (L.)

Rush skeletonweed
Asteraceae
(Sunflower family)

Rush skeletonweed is a perennial, 1 - 4 feet tall. Stems have downwardly bent, coarse hairs; basal 4 - 6 inches, smooth above, erect, stiff, much branched with bare appearance. Leaves form in a basal rosette, sharply toothed, and withers as the flower stem develops. Leaves of the stem are inconspicuous, narrow and entire. Flowering heads are scattered on branches, approximately 3/4 inch in diameter, 7 - 15 flowered, yellow, strap-shaped. Fruits are pale brown to nearly black, about 1/8 inch long. Body of fruit is several-ribbed, smooth below with tiny scaley projections above, terminated by numerous soft white bristles.

Rush skeletonweed is an introduced Eurasian species which presently infests several million acres in Idaho, Oregon, Washington and California. It generally inhabits well-drained, light-textured soils along roadsides, in rangelands, grain fields and pastures. Soil disturbance aids establishment. The extensive and deep root system makes skeletonweed difficult to control. Cut surfaces of the leaves and stems exude a milky latex. Flowering and seed production occur from mid-July through frost.

Rosettes resemble those of a dandelion, but are readily distinguished when stem elongation starts in early summer. Red colored hairs are present at the stem base at that time.

Yellow flowers, less than one inch across, have strap-shaped petals that are flat across the end with distinct lobes or teeth.

Horseweed
Conyza canadensis (L.) Cronq.

Horseweed

Asteraceae
(Sunflower family)

Horseweed is a winter annual or annual, 1 - 5 feet tall. Stems are erect, unbranched below but often branched above. Leaves are alternate, crowded on the stem, simple, bristley with hairs and sessile or short petioled. Lower leaves are spatulate and sparingly or coarsely toothed, while upper leaves are lance-shaped to linear. The inflorescence is branching with slender flower stalks, flowers are inconspicuous with small white ray and yellow disk flowers. Seeds are numerous, very small, flattened, about 1/16 inch long, with a white bristley pappus.

A native of North American grassland, horseweed is common in pastures, meadows, cultivated fields, along roadsides and in waste areas. Mowing infested meadows or pastures when the plants are in the bud stage will prevent seed productions. The leaves and flowers contain a terpene which is particularly irritating in the nostrils of horses. Flowering and seed production occur from late June to September.

Seedlings have stiff hair covering the narrow leaves that are individually attached alternately around the stem.

Small, inconspicuous white or yellow flowers are borne at stem terminals in mid-summer on this annual.

Curlycup gumweed
Grindelia squarrosa (Pursh) Dunal

Curlycup gumweed

Asteraceae
(Sunflower family)

Curlycup gumweed is a biennial or short-lived perennial that reproduces by seed and grows 1 - 3 feet tall. The root is fiberous; leaves alternate, 1 - 3 inches long, with saw-toothed margins. They are gland-dotted and exude a sticky material. Flowers are bright yellow, one inch in diameter, borne singly on the end of the branches. Curved bracts surrounding the flower also secrete a sticky substance which gives the plant its name. Seeds are oblong, cream colored, four-angled and deeply ridged.

Curlycup gumweed is a native plant that may be found in pastures, rangelands, roadsides and waste areas sometimes forming nearly pure stands. It is highly drought resistant and increases after periods of dryness. It is considered undesirable as forage and is unpalatable to livestock. Indians used gumweed as a treatment for asthma, bronchitis, colic and skin rash. Extracts of gumweed are used in todays medicine for treatment of bronchial spasm, asthma, whooping cough and poison-ivy rashes.

Leaves are alternate and rounded with a toothed margin and are also covered with a sticky resin.

Waxy yellow flowers appear in mid-summer on curlycup gumweed. The involucre surrounding the flower is covered with curved bracts which secrete a sticky resin.

Broom snakeweed
Gutierrezia sarothrae (Pursh) Britt. & Rusby

Broom snakeweed
Asteraceae
(Sunflower family)

Broom snakeweed is a native perennial warm-season plant reproducing by seed and root systems. The plant has a woody nature but rarely grows to heights over 18 inches. Stems grow new from the crown each year, not from old growth. Stems are close together, stiff and somewhat resinous. Leaves are alternate, narrow and untwisted, 1 - 1 1/2 inches long and are green with smooth margins, slightly rolled to the center and are located on the upper parts of the plant. Flowers are small and round with inconspicuous yellow ray flowers. Fruit are oval and covered with chaffy scales. Blooms appear from July to September with seed maturity taking place in September and October.

Broom snakeweed is a common poisonous plant which is most toxic during leaf formation. Most losses occur in cattle and sheep as abortions. Weak calves and lambs are common during times when pasture is scarce and broom snakeweed is eaten. Pastures are not properly utilized when this weed is found within the grasses. Control of this species can be attained with herbicides.

Spring growth of this perennial comes from the plant crown each year. Leaves are 1 - 1 1/2 inches long and are straight.

Small yellow flowers appear in mid-summer on this perennial. Herbicides are most effective when applied in the bud or early bloom stage.

Common sunflower
Helianthus annuus (L.)

Common sunflower and Nuttal sunflower

Asteraceae
(Sunflower family)

The common sunflower is an annual, 1 - 10 feet tall. Stems are erect, simple to much branching and rough. Leaves alternate and are simple, rough, hairy, ovate or heart-shaped, with toothed edges. The flower heads are showy, with yellow to orange-yellow ray flowers and brown or dark reddish-brown disk flowers. Achenes are 1/4 inch long, wedge-shaped, somewhat flattened to four-angled, smooth except for a few short hairs at the tip and gray to brown in color.

The nuttal sunflower, *Helianthus nuttallii* (T. & G.) is a perennial, 2 - 10 feet tall. Stems grow erect, arising from a cluster of enlarged, tuberous roots with short, thick rhizomes, smooth to rough. Leaves are opposite or sometimes nearly all alternate, ovate to narrowly lanceolate, entire or serrated edges. The flower heads are few and at the end of long peduncles. Ray and disk flowers are both yellow; achenes are 1/8 inch long.

Native to North America, the sunflower has been cultivated since pre-Columbian times for its edible seeds. They are common weeds of roadsides, fence rows, fields, pastures, waste areas and anywhere else conducive for growth. Flowering is from July to September.

Seedling plants of common sunflower have a rough surface with a slightly lobed margin tapered to a point.

Clusters of flowers appear in late summer on Nuttal sunflower. Flowers, approximately 2 inches across, are abundant.

Poverty sumpweed
Iva axillaris (Pursh)

Poverty sumpweed
Asteraceae
(Sunflower family)

Poverty sumpweed is a perennial, 6 - 18 inches tall. Stems are erect, much branched, smooth or slightly hairy. Leaves are mainly opposite, numerous, sessile, entire, narrowly oblong, 1/4-3/4 inch long, rough-hairy, harsh and stiff to the touch. The small tubular flowers are borne in small heads which hang from the axils of the upper leaves; achenes are deep gray to almost black, wedge-shaped, 1/8 inch long with a rough surface.

Native of western U.S., poverty sumpweed can be found from the Canadian border to Mexico. In Wyoming and Utah it commonly occurs along railroad rights-of-way and roadsides, in pastures and waste areas, and sometimes in cultivated cropland. It is quite tolerant of saline and alkaline soils.

Seedling plants have oval, rough leaves that appear as opposite pairs.

Small, tubular yellow flowers borne in leaf axils are produced in late summer on this perennial.

Marshelder
Iva xanthifolia (Nutt.)

Marshelder

Asteraceae
(Sunflower family)

Marshelder is a robust annual, 2 - 7 feet high, reproducing by seed, branching upward. Leaves are mostly opposite; but the uppermost alternate, with long petioles. Blade is ovate to broadly ovate, 3 - 12 inches long and simlar to those of common cocklebur. Flowering heads are similar to those of povertyweed; but they are stalkless, crowded on long, branching spikes at the top of the stem and the upper leaf axils. Flowers are greenish-white, with 8 - 20 staminate flowers and usually five pistillate flowers. Corollas are greatly reduced. Fruits are ovate, less than 1/4 inch long and dark brown.

Marshelder is an occasional weed in sandy, damp or drying sites, especially streambeds and flood plains. The pollen may cause serious hay fever in late summer, and the leaves produce a skin rash in some people. Flowering occurs from late July through September. It is native to middle North America.

Leaves, often over six inches across, have toothed margins with distinct veins on the surface. Opposite leaves are connected to the main stem with a long petiole.

Very small, white flowers and seeds appear as a mealy substance surrounding the terminal stems.

Prickly lettuce
Lactuca serriola (L.)

Prickly lettuce
Asteraceae
(Sunflower family)

Pricky lettuce is biennial or winter annual, from a large taproot, with milky juice, reproducing only by seeds. Principal stem is 1 - 5 feet tall, branching only in flowering portion, or sometimes a few short branches from the base. Leaves alternate, twisted at the base to lie in a vertical plane, prickly on the lower side of midrib, clasping the stem with two angled or earlike lobes. Lower leaves are 2 - 10 inches long with two forms. Margins may be pinnately lobed or lobeless. Flowering heads are numerous with 13 - 27 flowers. They are yellow, often drying blue, 1/8 - 1/3 inch broad and composed of ray flowers only. Each flowering head has six to 30 flattened fruits with five to seven parallel ridges on each side. They are short, bristly near the summit.

Prickly lettuce is a native of Europe, and is now naturalized throughout most of the U.S. This weed, is a serious invader in disturbed soil of irrigated crops and orchards. It is also common along roadsides, yards and small gardens. Hybrids may represent an infusion of genes from cultivated lettuce. Flowering occurs from July to September.

Leaves are slightly lobed in the seedling stage with sharp spines found on the back side of the midrib.

Small yellow flowers less than 1/3 inch across are numerous on the terminal branches of this annual.

Skeletonweed
Lygodesmia juncea (Pursh) D. Don

Skeletonweed

Asteraceae
(Sunflower family)

Skeletonweed is a perennial reproducing by seed. The plant growing to heights up to 18 inches appears to be all stems. The stems are erect green, finely grooved and branched. Leaves are small linear and up to two inches long at the base. Upper leaves are reduced to small awl-like projections. Stems and leaves contain a white, milky, latex sap. Light purple flowers are small--less than one inch across--and are flattened, but lobed at the tip. Seeds are less than 1/2 inch long, have eight to 10 ridges vertically and have a cream-colored papus. Blooms appear in late June through September with seeds maturing in July to September.

Skeletonweed is common along roadsides, waste areas, pastures, rangeland and in cropland. It is not controlled by most herbicides; therefore, it appears to be increasing in density in many areas. This plant could become a potential problem in cropland, so prevention would be advisable.

Upper leaves on skeletonweed are reduced to small awl-like projections.

Skeletonweed produces light purple flowers in late summer. Petals have rounded teeth on the tips.

Scotch thistle
Onopordum acanthium (L.)

Scotch thistle

Asteraceae
(Sunflower family)

Scotch thistle is a biennial that grows up to 8 feet tall. Stems have broad, spiny wings. Leaves are large, spiny, and covered with fine dense hair giving a grayish appearance. They are alternate, coarsely lobed; basal leaves may be up to 2 feet long and 1 foot wide. Flower heads are numerous, 1 - 2 inches in diameter, bracts spine-tipped. Flowers are violet to reddish. Fruits are about 3/16 inch long, tipped with slender bristles.

Scotch thistle is native of Europe and eastern Asia and is now sparsely naturalized over much of the U.S. It can be found along waste areas and roadsides. It is an aggressive plant and may form stands so dense that they are inpenetrable to livestock. Scotch thistle can be controlled with herbicides.

Rosettes two feet across are not uncommon for this hearty plant. Leaves are covered with white hair giving them a blue-green color.

Purplish-pink flowers, 1 - 2 inches in diameter, are produced the second year of this biennial's growth cycle. Stems appear to have wings.

Tansy ragwort
Senecio jacobaea (L.)

Tansy ragwort
Asteraceae
(Sunflower family)

Tansy ragwort is a biennial or short-lived perennial from a small to medium taproot. Stems are 1 - 6 feet tall, solitary or several, simple up to the inflorescence. Their pubescence consists of cobwebby hairs in early stages of growth. leaves 2 - 8 inches long, alternate and equally distributed, mostly 2 - 3 times pinnately lobed, the terminal lobe generally larger than the lateral ones. Flowering heads are numerous. Both ray and disk flowers are yellow; ray flowers are 10 - 13 in number and 1/4 - 1/2 inch long. Fruits of the disk flowers are minutely pubescent; those of the ray flowers are glabrous.

This European native is widespread in Washington, Oregon and California, having arrived in seaports in the early 1900s. Tansy ragwort is toxic to cattle and horses. Like common groundsel, it has several alkaloids which produce irreversible liver damage. Flowering occurs from July to September. While tansy ragwort is not presently known in Utah or Wyoming, it does infest millions of acres of private range, public range and pasture land in the Pacific Northwest and may invade our states in the future.

Rosettes of this biennial have leaves that are deeply lobed--compound leaflets are also lobed.

Yellow terminal flowers less that 1/2 inch across are borne as terminal clusters on tansy ragwort.

Common groundsel
Senecio vulgaris (L.)

Common groundsel
Asteraceae
(Sunflower family)

Common groundsel is a simple or branched annual, sometimes biennial, from a taproot. Stems are 6 - 18 inches tall; herbage is glabrous to crisp-hairy. Leaves are alternate, coarsely and irregularly toothed or pinnately parted. Basal leaves are usually purplish on the under surface, 1 - 4 inches long and 1/2 to 1 1/2 inches wide. Heads are several to numerous, with yellow disk flowers; involucral bracts are black-tipped. Fruits are slender, ridged, somewhat pubescent, long, silky. White hairs terminate the achene. The genus *Senecio* is one of the largest genera of plants, containing well over 1000 species.

Common groudsel is native of the Old World, but is now widely distributed throughout most of the temperate zone. Flowering occurs from early spring to killing frosts. This weed is found on roadsides, fields and disturbed sites. The herbage is toxic to livestock. Toxicity is caused by several alkaloids which produce irreversible liver damage, much the same as tansy ragwort.

Seedling plants appear as tiny rosettes with sharply notched leaves. Plants contain alkaloids poisonous to cattle and horses.

Flowers appear as yellow buttons with no ray flowers present, then burst into a circular bundle of windblown seed. Flower and seed production start in early summer and continue until frost.

Canada goldenrod
Solidago canadensis (L.)

Canada goldenrod
Asteraceae
(Sunflower family)

Canada goldenrod is a perennial growing to heights of 4 feet, spreading by creeping roots and seeds. Alternate leaves surround the central stems and gradually reduces upwards. Leaf blades are entire or commonly toothed having three veins. Flowers are yellow, borne on numerous small heads with overlapping involucral bracts. Thirteen ray flowers are common, but they may contain as few as 10 or as many as 17. Fruit are one seeded, tipped with a circle of white hairs. Solidago is a genus of nearly 100 species, several common in Wyoming and Utah.

Canada goldenrod, a native to North America, is common in the U.S. and Canada. The plant is often seen along roadsides, streambanks and along ditches. The extensive root system makes it a difficult plant to control. It seldom reaches densities that are a problem in rangeland. Flowering occurs from August to October.

Leaves are alternately spaced, 3 - 5 inches in length. They have three distinct veins and are sometimes toothed.

Golden yellow flowers grow in clusters on the terminal stem in late summer.

Marsh sowthistle
Sonchus arvensis (L.) ssp. *uliginosus* (Bieb.) Nyman

Marsh sowthistle

Asteraceae

(Sunflower family)

This perennial arises from an extensive creeping root system, commonly 1 - 6 feet tall, glabrous, at least below the inflorescence and often covered with a waxy bloom which can be rubbed off. Herbage has a milky juice. Leaves prickly-margined, the lower and middle ones are usually pinnately lobed to pinnatifid, mostly 4 - 10 inches long and 1 - 1 1/2 inches wide. Upper leaves are progressively less lobed and become clasping, uppermost leaves bractlike. Flowering heads grow in a terminal flat to a domelike cluster, 1 - 1 1/2 inches across at anthesis. It has yellow ray flowers, which are fertile. Achenes are flattened, ribbed and rugose, with pappus of capillary bristles.

Marsh sowthistle is native to Eurasia, but is is now found worldwide in temperate zones. Two subspecies are recognized; our material belongs to subspecies *uliginosus*. Plants occupy arable land, riverbanks and fields usually on nitrogenous loamy soils. It is a moderate to vigorous user of nitrogen and competitor for space. Hoeing favors vegetative dissemination. Flowering occurs from July to September.

Leaves have sharp prickles on their lobed margins. Immature growth stages often come from extensive underground roots.

Yellow flowers appear in late summer, they develop into a cluster of windblown seeds. Just before flowering, buds have large bottoms, but taper to the top.

Common tansy
Tanacetum vulgare (L.)

Common tansy
Asteraceae
(Sunflower family)

Common tansy is an aromatic perennial. Stems are 1 1/2 - 6 feet tall. It reproduces from seeds and rootstalks. Leaves are alternate, deeply divided into numerous, narrow, toothed segments. Flowers are numerous in flat-topped dense clusters. They are yellow, 1/4 - 1/2 inch across. Seeds are yellowish-brown with short five-toothed crowns.

Common tansy is native of Europe and became established in the U.S. when introduced as an ornamental. It is generally found along roadsides, waste areas, streambanks, and in pastures throughout most of the U.S. It is undesirable as forage for livestock. However, it has long been used as a medicinal herb. Common tansy is sometimes confused with tansy ragwort.

Leaves are divided into individual leaflets, which are serrated on the margins. Stems are often purplish-red in color.

Showy yellow, button-like flowers without petals explain why this plant was originally introduced as an ornamental.

Dandelion
Taraxacum spp.

Dandelion
Asteraceae
(Sunflower family)

Dandelion is a perennial herb from a long, often branched taproot up to 6 inches long, with milky juices. Aerial stems are lacking, reproducing by seeds and by new shoots from the root crowns. Leaves are clustered at the top of the root crown. They vary in size, from 2 - 12 inches long, divided into pairs of lobes, which are pointed or blunt at the tips. Flowering heads are 1 - 2 inches across, composed of yellow, petal-like, ray flowers. Heads are solitary at the end of long, hollow stalks, up to 2 feet long. Achenes are 1/8 inch long, five- to eight-ribbed, apex ends in a slender beak two or four times as long as the body of the achene, parachute-like hairs at apex of beak. The genus *Taraxacum* is a taxonomically confusing group subject to many interpretations as to the number of species. In fact, hundreds of specific names have been published. Most authorities now favor lumping the taxa into two species--*T. laevigatum* and *T. officinale*.

The common dandelion is native to Europe, but is now cosmopolitan. It grows in moist sites, including lawns, meadows, pastures and overgrazed areas. It is good forage on the ranges, and is especially relished by sheep and cattle. Flowering occurs almost nine months of the year.

Seedlings and immature growth stages of this perennial have leaves that vary in size from 2 - 12 inches long and are usually deeply lobed.

Yellow flowers on this perennial develop into a showy, circular ball of windblown seeds.

Western salsify
Tragopogon dubius (Scop.)

Western salsify
Asteraceae
(Sunflower family)

Western salsify is biennial or sometimes annual, arising from a long taproot several inches long. Stem is 1 - 3 feet tall, more or less branched. Herbage, has milky juice. Some pubescence occurs when young, but it is glabrous at maturity. Leaves are long and narrow, up to 12 inches long, gradually tapering from the base to the apex. Flowering heads occur at the end of long, hollow peduncles; involucral bracts are usually about 13, or only eight in depauperate plants, 1 - 2 inches at anthesis. They distinctly surpass the pale, lemon-yellow ray flowers as the fruits mature. Achenes with a 1 - 1 1/2 inch slender beak at the apex. The beak is topped by a pappus of whitish plumose bristles.

Western salisfy is native to Eurasia and is now established over much of temperate North America. This is a weed on roadsides and disturbed waste sites, especially along the edges of grain fields. Two other species of this genus are also in our states, and sometimes all three species hybridize with each other. Because the involucre and receptacle reflex at maturity, the achenes form a characteristic "ball" of fruits, similar to that of the common dandelion. Flowering occurs from May through July.

Seedling and immature stages of western salsify have long, narrow leaves containing a white milky latex sap.

Flowers start to appear in mid-summer. They have a pale yellow color with distinctive involucral, green bracts extending past the petals.

Common cocklebur
Xanthium strumarium (L.)

Common cocklebur

Asteraceae
(Sunflower family)

Common cocklebur is an annual, 2 - 4 feet tall, stem erect, branched, ridged, spotted and very rough, leaves alternate, triangular or heart-shaped, rough on both sides and long petioled. The flowers are small, in axils of upper leaves; male and female flowers are separate. Fruits are 1 inch long, woody, with hooked prickles and two curved spines at tip. Two seeds appear per pod. They are flattened and pointed on tips, dark brown.

Common cocklebur is native to North America, but is now worldwide in distribution. Several other species are present in Wyoming and Utah. Cocklebur is common in cultivated fields, abandoned land, run down pastures, road ditches and waste areas. The burs are irritating both to humans and animals, and the burs in wool depreciate its value. Both the seeds and seedlings contain a poisonous substance which is toxic to livestock. Flowering may occur from July to September.

Seedling plants have very rough leaves that are notched on the margins and taper to a point. Careful removal of seedlings revealing the old bur from which it came allows for positive identification.

Large rough leaves are attached to a thick purplish-green stem with distinctive purple to black spots along the stem

Woodyaster
Xylorhiza glabriuscula (Nutt.)

Woodyaster
Asteraceae
(Sunflower family)

Woodyaster is a perennial reproducing from seeds. The plants are less than 12 inches in height with many stems arising from a single crown and root. Stems are unbranched, have alternate lance-shaped, entire leaves which produce a daisy-like flower on the top of each stem. White flower petals surround a dark central disk, leaves are hairy with smooth margins and are rounded at the tip. Seeds are numerous, elongated and are tipped by a brown tuft of hairs. Roots are extensive taproots, woody in nature.

Woodyaster is native to Wyoming and Utah and is an indicator of selenium soils. Livestock can be poisoned by this plant if consumed over extended periods. Decomposition of the woodyaster plants leaves selenium in a form readily absorbed by surrounding desirable vegetation.

Roots of this perennial are very large and woody. Plants grow to heights under 12 inches.

Flowers appear in early summer resembling small daisies with white ray flowers and yellow disk flowers in their center.

Bur oak
Quercus macrocarpa (Michx.)

Bur oak

Fagaceae
(Oak family)

Bur oak is a perennial, coarse shrub to large tree with dark, deeply furrowed bark. Leaves are variable, sometimes very large (10 inches long), usually deeply lobed, dark green on top, silvery and hairy below. Fruits mature the first year; the cup holding the acorn is deep. The margin and often the sides, are distinctly fringed; the margin covers 1/2 - 2/3 or more of the acorn. The acorn is egg-shaped, up to 1 1/2 inches long.

Bur oak is a native that may be a shrub or reach heights of 20 feet or more. Most species of oak can be a problem because they contain toxic substances; however, Gambel and shinnery oak are responsible for most livestock losses. Cattle may be poisoned by browsing on young oak, sprouts in cutover areas, leaves and acorns. Oak is most toxic in budding and leafing stages. Toxicity decreases as leaves mature. Cattle can tolerate up to 50 percent of oak browse in their diet without showing ill effects. However, more than 50 percent oak browse in the diet will cause sickness and more than 75 percent can cause death. Cattle losses can be reduced by keeping animals off oak ranges until other forage is available and the oak leaves mature. Chemical control of oak is difficult.

Bur oak leaves contain tannins causing cattle poisoning. Leaves are 4 - 8 inches long with rounded lobes.

Acorns of bur oak are borne in deep cups that are fringed at the top during late summer and fall.

Western snowberry
Symphoricarpos occidentalis (Hook.)

Western snowberry
Caprifoliaceae
(Honeysuckle family)

Western snowberry is a perennial shrub, 1 - 3 feet tall. It reproduces by seeds and offshoots from rootstalks. Branches are light colored, upright or somewhat spreading. Leaves are simple, ovate, sometimes with several large blunt teeth. They are opposite, 3/4 - 2 3/8 inches long, 3/8 - 1 3/8 inches wide. Flowers are bell-shaped and pinkish. Fruits are round, 3/8 inches wide, shiny, white (drying blackish or bluish). They persist throughout the winter.

Western snowberry is a native shrub common to hillsides, open woods, river banks and low grasslands. It can form dense thickets which crowd out more desirable forage, however, snowberry is rated as fair forage for cattle and sheep and may provide fair to good forage for deer and elk. The berries also provide winter feed for upland birds.

Pinkish-white flowers, bell-shaped, appear in mid-summer and continue through fall.

Clusters of white berries that are 3/8 inch long appear in late summer on reddish stems in axils of oval-shaped leaves.

Chokecherry
Prunus virginiana (L.)

Chokecherry
Rosaceae
(Rose family)

Chokecherry is a shrub or tree, 4 - 25 feet tall, bark is gray to brown, smooth to somewhat scaly. Leaves are alternate, usually with conspicuous glands at the apex of the petiole. The leaf is 1 - 3 inches long and 1 - 1 1/2 inches wide, abruptly acute at the apex, with serrate margins. Flowers are numerous in 2 - 3 inch long racemes that appear after leaves expand. Petals are white; stamens are numerous. Fruits look like miniature cultivated cherries, red at first and becoming black at maturity. Insect galls often replace developing fruits.

Chokecherry grows along streams and in open forests in both Utah and Wyoming. The leaves, bark and seeds of this shrub contain the glycoside amygdalin, which hydrolyzes to highly toxic hydrocyanic acid. The ripe fruit is edible, provided the seeds are first removed, and it is used for making jams and jellies. Children have been poisoned by chewing the leaves or seeds. Cyanide poisoning can occur very quickly and with deadly results. Flowering occurs in May and June.

Clusters of small white flowers with yellow centers appear in early spring on this woody perennial. Leaves contain hydrocyanic or prussic acid, which are highly poisonous.

Edible purplish-black berries are formed in clusters along all branches and stems in late summer.

Poison-ivy
Toxicodendron rydbergii (ex Rydb.) Greene

Poison-ivy

Anacardiaceae
(Cashew family)

Poison ivy is a perennial sub-shrub often less than 3 feet tall, to a distinct shrub up to 9 feet tall. Stems are simple or sparsely branched, arising from much branched subterranean stolons. Aerial roots are absent, plants never climb. Leaves alternate, compound usually with 3 leaflets. Flowers grow in axillary panicles or racemes, with five sepals and united at the base. They have five greenish petals. Three-parted stigmas turn black with age. Fruit is globose, cream to yellow, 1/4 inch broad, wrinkled at maturity and glabrous.

This plant occupies many habitats, including floodplains, river terraces, railroad rights-of-way and other disturbed sites. Seed dispersal by birds and mammals and greater environmental disturbance indicate that poison-ivy will continue to occupy new sites. The poison in this plant is a milky oil found in the phloem. Since the poison occurs in ducts within the plant, the plant must be damaged in some way to expose the poison ducts before the poison can be transmitted. If the liquid oil or its black, dried residue, comes in contact with the skin, a debilitating rash will follow in a few hours. The poison can also be transmitted through secondary objects such as tools.

Leaves are often purplish in early spring growth and late fall. They usually have three leaflets per leaf, an important identification characteristic of this perennial.

Fruit with ridges resembling tiny green pumpkins are formed on this poisonous species in late summer.

Jointed goatgrass
Aegilops cylindrica (Host.)

Jointed goatgrass
Poaceae
(Grass family)

Jointed goatgrass is a winter annual, 15 - 30 inches tall with erect stems that branch at the base giving the plant a tufted appearance. Leaves are alternate, simple with auricles at the base and a leaf blade 1/8 - 1/4 inch wide with hairs. The spike is cylindric, more than 10 times as long as it is wide. It contains two to 12 spikelets which fit into the contour of the rachis, spikelets 1/2 inch long with one to three viable seeds. Glumes several ribbed with a keel on one side extending into a single awn or beard. At maturity the spike falls intact and spikelets separate with a segment of the rachis still attached.

Jointed goatgrass is native to southern Europe; but it is now established in most winter wheat growing areas of North America, introduced many times as a seed contaminant or by custom combiners. It is found mostly in wheat fields, but is survives along roadsides, in waste areas, alfalfa fields and pastures. The plant is most difficult to control in areas where winter wheat is grown continuously. Flowering and seed production may occur from May to July.

Immature jointed goatgrass plants have marginal hairs on the leaf and at the stem juncture.

Segmented, slender jointed goatgrass compared to wheat at harvest time.

Quackgrass
Agropyron repens (L.) Beauv.

Quackgrass
Poaceae
(Grass family)

Quackgrass is an aggressive perennial grass reproducing by seed, or spreading by a shallow mass of long, slender, branching rhizomes. Rhizomes are usually yellowish-white, sharp-pointed, somewhat fleshy. They are able to penetrate hard soils or even tubers and roots of other plants. Stems are erect and usually 1 - 3 feet tall. Leaf blades are 1/4 - 1/2 inch wide, flat, pointed and they have small auricles (ear-like appendages) at the junction of blade and sheath. Leaf sheaths and the upper surface of leaf blades may be thinly covered with soft hairs. Spikelets are arranged in two long rows, borne flat-wise to the stem. Florets are awnless, or with short straight awns.

Quackgrass is an introduced species, originating in the Mediterranean area. It has spread over much of North America, adapting well to moist soils in cool temperate climates. Quackgrass reduces productivity in crops, rangeland and pasture. It is also a serious nusiance in lawns, ornamentals and home gardens. It is a highly competitive weed, and is believed to be allelopathic. Because of the ability of broken rhizome segments to grow and produce new plants, it is extremely difficult to control mechanically.

New plants develop in dense stands from sharp pointed yellow-white branching rhizomes.

Spikelets are borne in two rows flat-wise to the stem. Pollinating florets are either beardless or have short, straight beards (awns).

Wild oats
Avena fatua (L.)

Wild oats
Poaceae
(Grass family)

Wild oats is an annual, 1 - 4 feet tall with erect hollow stems. Leaf blades are 1/8 - 5/8 inch wide, sheaths open, ligules membraneous, seedling leaves twist counter clockwise. The inflorescence is an open panicle, 4 - 18 inches long; drooping, spikelets contain 2 - 3 florets which disarticulate above the glumes. Seeds are yellow to black, narrowly oval, 1/4 - 1/2 inch long. This species is distinguished from tame oats by the twisted awn which bends at right angles and a horseshoe shaped scar at its seed base.

Wild oats is a native to Europe, but is now common throughout much of western North America. It is a serious problem in spring seeded small grain; but it also occurs along roadsides, in pastures and waste areas. Seed can remain dormant in the soil for as long as 10 years, making it difficult to eliminate once established. Flowering and seed production occur from June to August.

Seedling leaves twist counter clockwise when viewed from the plant stem. Small marginal hairs can be seen on leaf edges.

Spikelets contain two or three seeds which have awns that form right angles at maturity.

Japanese brome
Bromus japonicus (Thunb.)

Japanese brome
Poaceae
(Grass family)

Japanese brome is an annual or winter annual, 14 - 30 inches tall. It reproduces by seed. Leaf blades and sheaths are soft and hairy; inflorescence diffuse 4 1/2 - 8 inches long with 3 - 5 lower, usually drooping branches. Spikelets are broad and about 1/2 inch long, awns are 1/4 - 3/4 inch long, somewhat twisted and widely spread at maturity.

This European introduction has become a nuisance in depleted ranges, hayfields, and dry soils in waste or disturbed areas. It is often confused with downy brome. Seed germination usually occurs in the fall; young plants overwinter, grow rapidly in the spring and reach maturity by June. It occurs throughout the West and is only palatable in the early stages of growth before seeds dry in the spring. A vigorous cover of desirable, perennial grasses is a good method of prevention in pasture and rangelands.

Stems and leaves are covered with soft hairs at all growth stages.

Mature spikelets about 1/2 inch long have awns from 1/4 - 3/4 inch long.

Downy brome
Bromus tectorum (L.)

Downy brome
Poaceae
(Grass family)

Downy brome is an annual or winter annual, 4 - 30 inches tall; it reproduces by seed. Leaf sheaths and blades are densely covered with soft hair, flat blades. Ligules are very short. Inflorescence is dense, slender, usually drooping. One-sided, 2 - 6 inches long. Spikelets are nodding, slender, 3/8 - 3/4 inch long. Awns are 3/8 - 5/8 inch long, usually purplish at maturity.

Downy brome was introduced from the Mediterranean region in packing material and first found near Denver, Colorado. An indicator of over-grazing, it is now widely distributed throughout North America and is common along roadsides, waste areas, misused pastures and rangelands, and cultivated crop areas. Although downy brome is considered an invader, on certain intermountain ranges it has become the primary green forage utilized by livestock. The plant heads out in early June and competes with more desirable perennial grasses for moisture. After it matures it becomes a nuisance and a fire hazard. It is also a common crop seed contaminant, especially abundant in crested wheatgrass and bromegrass.

Immature plants are covered with soft hairs at all growth stages.

Mature downy brome seed heads, 2 - 6 inches long, contain seed. They are 3/8 - 3/4 inch long, with awns up to 5/8 inch long.

Longspine sandbur
Cenchrus longispinus (Hack.) Fern.

Longspine sandbur

Poaceae
(Grass family)

This is an annual grass with tufted stems. It grows 8 inches to 3 feet tall, occasionally erect, but usually spreading horizontally and forming dense mats. Leaf sheaths are flattened, very loose, smooth with hairy margins. Leaf blades are flat, roughened, 2 - 6 inches long and 1/4 inch wide with rounded margins. The spikes are 1 - 3 inches long and bear clusters of 10 - 30 burs. Burs are thickly set with stiff, sharp, spreading spines. They usually contain two light brown, oval to oblong seeds.

A native of Europe, longspine sandbur is a nuisance throughout most of the U.S. It grows in cultivated fields, pastures and waste areas; but it favors sandy or well-drained, gravelly soils. It can be particularly troublesome to livestock causing injury to mouths, noses or eyes that come in contact with the mature burs. The presence of burs may also reduce the value of wool. Sandbur is commonly spread by animals and machinery. Flowering and seed production occurs from July to September.

Immature plants, often purplish, have flattened stems usually spreading horizontally.

Sandbur spikes are 1 - 3 inches long with 10 - 30 burs contained in each cluster.

Bermudagrass
Cynodon dactylon (L.) Pers.

Bermudagrass

Poaceae

(Grass family)

Bermudagrass is a warm-season, wiry perennial with long, slender, creeping rhizomes. Decumbent stems also spread laterally over the ground, rooting freely at any node. The flowering stems are upright and bear a terminal group of 4 - 5 spikes (a configuration resembling fingers on a hand.) Spikes are 1 - 2 inches long and bear two rows of sessile spikelets along one side of a somewhat flattened rachis. Leaves are generally smooth, with a conspicuous ring of white hairs at the junction of blade and sheath. Bermudagrass is sometimes confused with crabgrass (Digitaria sanguinalis, also a serious weed) because of superficial similarities between inflorescences, and the fact that both have spreading, stolon-like, rooting stems.

Bermudagrass is an introduced weed, probably originating in Eurasia. It has become widely established in warmer climates like southern Utah, where is is used as a pasture and lawn grass. It has now spread into some northern counties, and poses a serious threat to crop production and turf management. Bermudagrass can easily be spread by mechanical means.

Decumbent stems spread laterally over the ground, rooting freely at any node.

The flowering stems are upright and bear a terminal group of 4 or 5 spikes up to 2 inches long.

Barnyardgrass
Echinochloa crus-galli (L.) Beauv.

Barnyardgrass
Poaceae
(Grass family)

Barnyardgrass is a vigorous, warm-season, annual grass reaching 1 - 5 feet in height. Leaf blades are flat, broad (3/8 - 5/8 inches), smooth, and without a ligule or auricles at the junction of sheath and blade. Leaf sheaths are open, and blades are rolled in the bud. Panicles are often reddish to dark purple, as are the bases of many barnyardgrass stems. Spikelets are crowded; each one has a short, conspicuous awn and scattered stiff hairs.

Barnyardgrass was introduced from Europe and has become widespread throughout our region, especially in irrigated agronomic crops, gardens and other cultivated areas.

Seedling barnyardgrass plants, often purple at the base, have flattened stems.

Seed heads, often purple, contain crowded large seed in spikelets each containing a short, stiff awn.

Stinkgrass
Eragrostis cilianensis (All.) E. Mosher

Stinkgrass
Poaceae
(Grass family)

This is an annual having mostly open panicles with elongated multifloreted spikelets. Plants may range in height from 6 inches to 2 feet. Stems are hollow, sheaths open. Ligule consists of a fringe of straight hairs. Small glandular structures on the foliage and spikelets give off a disagreeable odor. Seed shatters readily at maturity, and appears as naked, egg-shaped, tiny grains (approximately 1/35 inch long).

Stinkgrass, introduced from the Old World, is found throughout the United States. It is common in waste places, gardens, alfalfa and other other crops. More than one species is reported to occur in our region.

Below the leaf at the stem juncture is a fringe of straight hairs.

Many egg-shaped seed are contained in these multifloreted spikelets of stinkgrass.

Foxtail barley
Hordeum jubatum (L.)

Foxtail barley and Squirreltail

Poaceae
(Grass family)

Foxtail barley is a perennial that reproduces by seed. Plants grow 1 - 2 feet tall and produce a pale green, bushy, open spike. At maturity these heads break into seven-awned clusters consisting of three spikelets (one fertile and two sterile). Spikelets are one-flowered. Awns are 1 - 2 1/2 inches long. Leaf sheaths may vary from smooth to densely hairy; leaf blades are 1/8 - 1/4 inch wide. Squirreltail, *Sitanion hystrix* (Nutt.) J.G. Sm., is a related perennial grass with open spikes and long, minutely-barbed awns. The most obvious difference between the two genera is the fact that *Sitanion* has two or more florets per spikelet, and two spikelets per node.

Foxtail barley is a native of North America, and is especially common in wet or alkaline, rundown meadows and pastures. Squirreltail is more common on dry hills, plains, open woods or rocky slopes. Both foxtail barley and squirreltail are palatable to livestock at younger stages; ;but awns of mature plants can cause serious injury to animals eyes, nose, throat and ears. At least three species of *Sitanion* and at least seven species of *Hordeum* (in addition to barley) are reported to occur in our region.

Immature plants are blue-green in color, and appear as clumps once established.

Squirreltail resembles foxtail, but squirreltail has open spikes with long, minutely-barbed awns.

Witchgrass
Panicum capillare (L.)

Witchgrass
Poaceae
(Grass family)

Witchgrass is an annual, 1 - 2 feet tall, reproducing by seed. Leaf blades are 1/8 - 9/16 inch wide. Spikelets form a large open panicle with many stiff branches, each bearing a single, small, shiny, green or grayish seed at the tip. Each plant is soft and hairy throughout.

Witchgrass is a common weed found along roadsides, waste places, cultivated fields, depleted hayland and ranges in poor condition. It may become a nuisance on cultivated fields with poorly established crops. It does not compete well with established perennial grasses. Therefore, it is not a serious problem on ranges in good condition. Witchgrass is somewhat palatable before seed heads develop, but is worthless once it matures and is damaging to hay quality. When mature the whole plant breaks away, scattering seed as it tumbles across the land.

All parts of this plant are extremely hairy. Leaves are often 1/2 inch wide on immature plants.

Spikelets form a large open panicle with many stiff branches, each bearing a single seed.

Wild proso millet
Panicum miliaceum (L.)

Wild proso millet
Poaceae
(Grass family)

Wild proso millet is an annual, 2 - 6 feet tall, with erect stems that branch at the base. Leaf blades are hairy, 1/2 - 3/4 inch wide; sheaths open with long spreading hairs. The ligule is a fringe of dense hairs 1/16 inch long which are fused at the base. The inflorescence is a spreading panicle 6 - 12 inches wide and not fully extended from the leaf sheath. Spikelets are two-flowered, the upper floret is fertile, the lower is sterile. Glumes are 1/8 inch long, ovate, pointed at the tip and strongly nerved. Seeds, shed at plant maturity, are smooth, shiny, olive brown to black in color. Seedling plants can often be identified by the attached seed coat on the roots.

The exact origin of wild proso millet is not known; however, it may have come from Asia or Central Europe. Wild proso millet is a prolific seed producer and a vigorous competitor in row crops. The seed can be spread by harvesting equipment, uncomposted manure, birds, small animals and irrigation water. Wild proso millet infestations often start at field entrances or along roadsides, so careful attention should be paid to these areas to prevent establishment. Flowering and seed production occur from July to September.

Seedlings are hairy, have wide leaves and can often be carefully removed from the soil to expose the black undeteriorated seed from which they came.

Seed color, ranging from olive brown to black, is the only positive identification characteristic distinguishing wild proso millet from domestic proso millet (which has yellow or light brown seed).

Yellow foxtail
Setaria glauca (L.) Beauv.

Yellow foxtail and Green foxtail
Poaceae
(Grass family)

Yellow foxtail is a tufted annual, 1 - 3 feet tall, erect stems that branch at the base, leaf sheath smooth, 2 - 8 inches long and 1/2 inch wide. The leaf blade is smooth, 1/8 - 3/8 inch with distinct hairs on leaf margin near the base. It is cylindrical with crowed spikelets, which are subtened by 6 - 10 long yellowish bristles. Seeds are broadly oval, green to yellow to dark brown, coarsely roughened and approximately 1/8 inch long.

Green foxtail, *Setaria viridis* (L.) Beauv., is generally shorter in height with roughened leaf sheaths without hairs.

These plants are native to Eurasia, but common throughout most of North America (green foxtail, is more common than yellow foxtail in the western U.S.). They are often serious problems in spring seeded alfalfa row crops and small grain crops, but they also occur along roadsides and waste areas. These plants are responsible for reductions in yield, increased cleaning costs, and expensive control measures. Flowering and seed production occur July to September.

Yellow foxtail and green foxtail look identical, except that yellow foxtail has numerous long hairs at the base of the leaf.

Green foxtail spike-like panicles range from 1 - 4 inches long with crowded spikelets.

Johnsongrass
Sorghum halepense (L.) Pers.

Johnsongrass
Poaceae
(Grass family)

Johnsongrass is a vigorous perennial plant resembling sudangrass. Erect stems are generally solid, have prominent nodes, and grow 2 - 8 feet tall. Leaf blades are flat with conspicuous midveins, and are often as much as 1 inch wide. Ligules are short and membraneous, with a terminal fringe of fine hairs. Mature inflorescence is a large open panicle bearing many awn-tipped, shiny, reddish to purple spikelets. Awns are bent and needlelike, and not all spikelets have awns. Plants spread by seed or by creeping, robust, fleshy rhizomes.

Johnsongrass is believed to have been introduced from the Mediterranean region as a hay or forage crop. It is now considered one of the world's worst weeds. Though once thought of as a warm-climate grass, it is now known to survive the cold winters of northern states. It is a very serious weed in Utah's Dixie, and is found in several of Utah's northern counties. It is extremely competitive and aggressive, and is especially troublesome in corn. Plants may form hydrocyanic acid, so under stress conditions, johnsongrass can be poisonous to livestock. Young plants are most toxic, especially when growth is interrupted by frost or drought.

Mature panicles or seed heads bear many awn-tipped, shiny, purplish spikelets.

A common method of propagation is from large fleshy rhizomes in addition to seed.

Medusahead
Taeniatherum caput-medusae (L.) Nevski

Medusahead

Poaceae

(Grass family)

Medusahead, introduced from Eurasia, is an aggressive winter annual with long-awned spikes that are nearly as wide as they are long. Plant height ranges from 6 - 24 inches. leaves are narrow (generally no more than 1/8 inch wide), and blades are more or less rolled. Mature awns are twisted, 1 - 4 inches long, stiff, and minutely barbed. Flowering and seed formation occur in May and June. Sometimes confused with foxtail barley or squirreltail, medusahead can be distinguished by the fact that its spike or head does not break apart as seeds mature. Instead, individual awned-florets fall away, leaving a bristly head made up of awn-like glumes that will persist over winter. Medusahead seedlings appear very similar to downy brome, except the latter is much hairier.

It is predominant on millions of acres of semi-arid rangeland in the Pacific Northwest. It is extremely competitive, crowding out even such undesirable species as downy brome. Infested ranches have suffered 40 - 75 percent reductions in grazing capacity. It has not been reported in Utah or Wyoming, but it should be considered a potential invader. Control of small isolated infestations is critical.

Medusahead is a highly competitive plant that crowds out all other vegetation on infested rangeland.

Twisted awns or beards are a good identification characteristic of medusahead.

Yellow nutsedge
Cyperus esculentus (L.)

Yellow nutsedge

Cyperaceae
(Sedge family)

Yellow nutsedge is an aggressive perennial species superficially resembling a grass. Plants range from 6 - 30 inches tall, with three- ranked leaves, and three-angled (triangular in cross section) pithy stems. True leaves originate from the base of each stem, while long leaf-like bracts radiate out from a common point just below the umbrella-like flower cluster. Otherwise, stems are naked. Leaves and stems have a waxy or shiny appearance. Spikelets are yellowish brown, and are borne on the ends of several to many slender branches of unequal length. Yellow nutsedge can spread by seed, creeping rootstocks, or by small underground nutlets. The many hard brown nutlets (1/2 - 3/4 inch long) may lay dormant in soil for some time before producing new plants.

Yellow nutsedge was probably introduced from the Old World, and has now invaded cultivated agricultural lands throughout North America. It prefers moist soils, becoming most troublesome in potatoes, beans, corn, gardens and ornamentals. *Cyperus* is genus of some 600 species, but yellow nutsedge is the primary weedy species in our region.

Immature plants have three-ranked leaves, a characteristic unlike grasses with which they are often confused.

Underground nutlets often propagate new plants in cultivated areas.

Seaside arrowgrass
Triglochin maritima (L.)

Seaside arrowgrass

Juneaginaceae

(Arrowgrass family)

Seaside arrowgrass is a native perennial, reproducing from rhizomes and seeds. Roots are short and fibrous. Basal leaves thick and succulent-like, having a round appearance but a rounded indention on one side. Seeds are borne on round stalks 1 - 4 feet high. Small, green flowers are inconspicuous with fruit borne in small pods having six compartments each containing a single seed. The pods surround the seed stalk, but each is attached by a short pedicel or stalk.

Arrowgrass grows in small patches in wet alkaline soils, starting growth in early spring with flowering and seed formation occurring in late summer. The plant contains hydrocyanic acid when drought or frost stressed, making it highly poisonous to livestock. As little as 1/4 pound of the stressed plants are capable of causing sudden death in yearling cattle if eaten in a short period of time. The poison, which causes respiratory failure, does not accumulate in the animal. Cured hay containing seaside arrowgrass has not been reported toxic to livestock.

Leaves of arrowgrass resemble grass leaves, but they are fleshy with a semi-circular groove down each leaf.

Fruit are individually connected to a main stem by a pedicel or short stem.

Wild onion
Allium textile (Nels. & Macbr.)

Wild onion

Liliaceae
(Lily family)

Wild onion is a native perennial reproducing by bulbs, aerial bulblets and seed. White flowers are borne on leafless flower stalks growing up to 18 inches in height. Several Allium species are common with flowers ranging in color from white to rose. Seeds are black with sharp angles 1/8 inch long or larger. Leaves are fleshy, and the basal has a strong onion odor. An underground bulb may be surrounded by smaller bulblets.

It is common in meadows and pastures and sometimes found in grain fields. The plant causes a strong onion flavor in milk and is difficult to control.

Underground bulbs are often surrounded by smaller bulblets.

White flowers, borne on leafless flower stalks, are surrounded by a thin membrane.

Great Plains yucca
Yucca glauca (Nutt.)

Great Plains yucca

Liliaceae
(Lily family)

Great Plains yucca is a perennial, 1 - 4 feet tall, and bush-like. It reproduces by seed and underground stems. Leaves are pale green 1 - 3 feet long, 1/4 - 1/2 inch wide, very stiff and sharply pointed, originating from the crown at ground level; flowers are large, greenish to creamy white, borne as a terminal cluster on a central stalk 1 - 4 feet tall; numerous seeds, black, flat, winged, 1/2 inch wide, are produced in an oblong fruit 2 - 3 inches long and about 1 inch thick.

Yucca is found throughout the West, mainly on dry sandy plains and prairies. Young plants and flowers are sometimes eaten by livestock. American Indians used the leaves to make baskets and the roots to produce soap lather. Pollination of yucca is dependent on a small insect, the yucca moth. The relationship between moth and plant is unique in that neither can complete its life cycle without the other.

Young plant's leaves end in a sharp spine and have thread-like tendrils at their base.

Large white or green flowers are borne along a central stalk that can be four feet tall.

Meadow deathcamas
Zigadenus venenosus (S. Wats)

Foothill deathcamas and Meadow deathcamas

Liliaceae
(Lily family)

Foothill deathcamas, *Zigadenus paniculatus* (Nutt.) S. Wats., and meadow deathcamas, native perennials have underground scaly bulbs and emerge in very early spring. Plants have five or six basal, thickened, unbranched V-creased leaves with a grass-like appearance. Plants reach heights of 2 feet. White to yellowish flowers are borne as a terminal cluster in early summer. Meadow deathcamas flowers are commonly branched with a panicle-like appearance while foothill deathcamas flowers form a terminal compact plume. The rough brown seeds are formed in a capsule.

All parts of the plant contain a poisonous alkaloid at all growth stages. Bulbs often mistaken for wild onions can cause severe illness in humans. Deathcamas is often eaten by sheep in early spring before other plants start producing forage. Sheep are commonly poisoned having respiratory problems after eating from 1/2 - 2 pounds of green forage. Pastures containing deathcamas should be sprayed in early spring and should not be grazed by sheep until late spring when more forage becomes available.

Leaves of immature plants have four or five unbranched V-creased leaves with a grasslike appearance.

Foothill deathcamas has a compact flower arrangement forming a single cluster rather than individual flowers up and down the terminal stem.

Glossary

Accrescent - Enlarging with age.

Achene - A small dry, one-seeded fruit which does not split at maturity.

Acute - Ending in a sharp point.

Adventitious - Not native; developing irregularly, accidentally or in unusual places.

Aerial - Growing in the air.

Areole - The spine-bearing area on a cactus.

Allelopathic Substances - Chemical compounds produced by plants that affect the interactions between different plants, including microorganisms.

Alternate - Singly along a stem; one leaf or bud at a node

Anthesis - Flowering; when pollination takes place.

Apex - The tip.

Appressed - Pressed flatly against the surface.

Ascending - Growing upward in an upcurved position.

Auriculate - Having small ear-shaped lobe or appendage.

Awn - A slender, usually terminal bristle.

Axil - The angle between a leaf and stem.

Axillary - Related to the angle between a leaf and stem.

Biennial - A plant which lives two years.

Bract - A small leaf-like structure surrounding a flower.

Calyx - The outer parts of a flower composed of leaf-like parts called sepals.

Capillary - Hairlike.

Caudex - The main axis of a plant including the stem and the root.

Chlorophyll - The green coloring matter of plants.

Clasping - Blade of the leaf extending beyond and surrounding the stem.

Clavate - Club-shaped.

Conical - Cone-shaped.

Cordate - Heart-shaped.

Corolla - The petals of a flower surrounding the stamens and pistil.

Crenate - With rounded teeth; scalloped.

Cylindrical - Having the form or shape of a cylinder.

Cyme - A flower cluster in which the central flower blossoms first.

Deciduous - Leaves falling at maturity or the end of a growing season.

Decompound - Divided into compound divisions.

Decumbent - Lying on the ground at the base, but rising at the tip.

Depauperate - Stunted.

Dimorphic - Occurring in two distinct forms.

Disk flower - A tubular flower in members of the sunflower family.

Elliptic - Oval or oblong with rounded ends.

Entire - Leaf margins that are not cut or toothed.

Filament - Anther-bearing stalk of a stamen; thread.

Floret - One of the closely clustered small flowers that make up the flower head of a composite flower.

Frond - The leaf of a fern.

Glabrous - Smooth; without hairs.

Glume - Bract at the base of grass spikelet.

Indehiscent - Not opening at maturity.

Inflorescence - The flowering part of the plant.

Involucre - A circle of bracts under a flower cluster.

Keel - A ridge; the two united front petals of a flower.

Lanceolate - Lance-shaped.

Leaflet - One small blade of a compound leaf.

Ligule - A thin, membranous outgrowth from the base of the blade of most grasses.

Linear - Long and narrow with parallel sides.

Lobed - Cut into shallow segments.

Membraneous - Thin and transparent.

Midrib - Central vein of a leaf.

Multifloreted - Many flowered.

Obicular - Circular.

Oblanceolate - Lance-shaped, tapering at both ends with the broadest part at tip end.

Obovoid - Egg-shaped with broader part at top.

Opposite - Arranged on the same node at the opposite side of the stem.

Orbicular - Circular

Ovary - The seed-bearing part of the pistil.

Ovate - Egg-shaped with broader part at base.

Palmate - Spreading like the fingers from the palm.

Panicle - Loose, irregularly compound flowering part of plant with flowers borne on individual stalks.

Pappus - Tuft of hair on the top of seed.

Pedicel - Stalk of a single flower.

Peduncle - Stalk of a flower cluster or individual solitary flower.

Perennial - A plant living more than two years.

Perfect flower - Having both stamens and pistils in the same flower.

Petiole - Stem or stalk of a leaf.

Phloem - The food transporting tissue of a plant.

Pinnate - Arranged along the sides of an axis.

Pinnatifid - Pinnately cleft.

Pistillate - Bearing pistils but no stamens.

Prostrate - Lying against the ground.

Pubescent - Covered with soft, short hairs.

Raceme - Arrangement of flowers along a stem on individual stalks about equal in length.

Rachis - The main stem bearing flowers or leaves.

Ray flower - Marginal petal-like flowers of some composites.

Receptacle - The part of the stem to which the flower is attached.

Reflexed - Turned abruptly downward.

Reticulate - Net-like.

Rhizomatous - Having rhizomes.

Rhizome - Underground stem, usually lateral, sending out shoots above ground and roots below.

Rosette - Cluster of compact leaves arranged in compact circle.

Rugose - Wrinkled.

Scale - Thin, dry membrane, usually foliar.

Sepal - One division of the calyx.

Serrate - Saw-toothed.

Sessile - Without a stalk.

Sheath - Lower part of the leaf which surrounds the stem.

Silicle - A short silique, almost equally as long as wide.

Silique - Enlongated capsule with a septum separating in two valves.

Simple leaves - Unbranched, not compound.

Spatulate - Spoon-shaped, narrow at base and wide at apex.

Spike - A dense raceme.

Spikelet - Small or secondary spike.

Spinose - Full of spines.

Stamen - The pollen-bearing organ of a flowering plant.

Staminate - Having stamens but no pistils.

Stellate - Star-shaped.

Stigma - Part of the pistil that receives the pollen.

Stolon - A horizontal stem which roots at the nodes.

Striate - Marked with lines or ridges.

Subtended - Underneath.

Succulent - Fleshy.

Terete - Circular in transverse section.

Toothed - Sawteeth-like projections on the margins of the blade.

Tuberous - Like a tuber or producing tubers.

Tufted - In compact clusters.

Umbel - Simple inflorescence of flowers on stalks which radiate from the same point or level.

Veins - Ribs of a leaf; vascular bundles of a leaf.

Winged - A thin, membraneous extension of a leaf blade.

References

Alley, H.P. and G.A. Lee. 1979. *Weeds of Wyoming*. University of Wyoming Agricultural Experiment Station Bulletin 498, Laramie. 120 pp.

Barkley, T.M. (ed.) 1986. *Flora of the Great Plains*. University of Kansas Press, Lawrence. 1392 pp.

Dennis, L.J. 1980. *Gilkey's Weeds of the Pacific Northwest*. Oregon State University press, Corvallis. 382 pp.

Dorn, R.D. 1977. *Manual of the Vascular Plants of Wyoming*. 2 vols. Garland Publishing. 1498 pp.

Gaines, X.M. and D.G. Swan. 1972. *Weeds of Eastern Washington and Adjacent Areas*. Camp-Na-Bor-Lee Assoc., Inc., Davenport, WA. 349 pp.

Gilkey, H.M. and L.J. Dennis. 1980. *Handbook of Northwestern Plants*. Oregon State University Press, Corvallis. 507 pp.

Goodrich, S. et al. 1986. *Uintah Basin Flora*. USDA Forest Service. 320 pp.

Hawkes, Robert B., Tom D.Whitson and LaRea J. Dennis. 1985. *A Guide to Selected Weeds of Oregon*. Oregon Department of Agriculture, Salem. 100 pp.

Hitchcock, C.L. et al. 1955. *Vascular Plants of the Pacific Northwest, Part 5: Compositae*. University of Washington Press, Seattle. 343 pp.

Holmgren, A.H. and B.A. Anderson. 1970. *Weeds of Utah*. Utah State University Agricultural Experiment Station Special Report 21, Logan. 115 pp.

James, L.F. et al. 1980. *Plants Poisonous to Livestock in the Western States*. USDA Agriculture Information Bulletin 415. 90 pp.

Johnson, J.R. and James T. Nichols. 1982. *Plants of South Dakota Grasslands - a Photographic Study*. SDSU Agricultural Experiment Station Bulletin 566, Brookings. 166 pp.

Nelson, E.W. and O. Burnside. 1979. *Nebraska Weeds*. Nebraska Department of Agriculture, Lincoln. 312 pp.

Robbins, W.W., M.K. Bellue, and W.S. Ball. 1970. *Weeds of California*. California Department of Agriculture, Sacramento. 547 pp.

Stubbendieck, J., S.L. Hatch and K.J. Kjar. 1981. *North American Range Plants.* Natural Resources Enterprises, Inc., Lincoln, Nebraska. 468 pp.

Wax, L.M., R.S. Fawcett and D. Isley. 1980. *Weeds of the North Central States.* USDA North Central Regional Research Publication 281. 303 pp.

Weed Science Society of America Standardized Plant Names Subcommittee. 1984. *Composite List of Weeds.* Weed Science Vol. 32, supplement 2, 1984. WSSA Champaign, Illinois. 137 pp.

Welsh, S.L. et al. 1987. A Utah Flora. Great Basin National Memoir #9. Brigham Young University. (In press)

Whitson, T.D., S.D. Miller and M.A. Ferrell. 1986. *Wyoming Weed Control Guide 1986.* University of Wyoming Agricultural Extension Service Bulletin 442R, Laramie. 95 pp.

Zimdahl, R.L. 1983. *Weeds of Colorado.* Colorado State University Cooperative Extension Service Bulletin 521A, Fort Collins. 220 pp.

Index

Notes

Notes